STUDY GUIDE

To Accompany
James T. McClave
and Frank H. Dietrich, II

STATISTICS

....

Sixth Edition

Susan L. Reiland

DELLEN
an imprint of
MACMILLAN COLLEGE PUBLISHING COMPANY
New York

MAXWELL MACMILLAN CANADA
Toronto

MAXWELL MACMILLAN INTERNATIONAL
New York Oxford Singapore Sydney

∙∙∙∙∙∙∙∙∙∙∙∙∙∙∙∙∙∙∙∙∙∙∙∙∙∙∙∙∙∙

© Copyright 1994 by Macmillan College Publishing Company, Inc. Dellen is an imprint of Macmillan College Publishing Company

Printed in the United States of America

All rights reserved. No part of this book may be reproduced or transmitted in any form or by any means, electronic or mechanical, including photocopying, recording, or any information storage and retrieval system, without permission in writing from the Publisher.

Macmillan Publishing Company
113 Sylvan Avenue, Englewood Cliffs, NJ 07632

ISBN 0-02-399210-7

Printing: 2 3 4 5 6 8 9   Year: 3 4 5 6 7

Contents

Preface		v
Chapter 1	What Is Statistics?	1
Chapter 2	Methods for Describing Sets of Data	4
Chapter 3	Probability	23
Chapter 4	Discrete Random Variables	46
Chapter 5	Continuous Random Variables	60
Chapter 6	Sampling Distributions	71
Chapter 7	Inferences Based on a Single Sample: Estimation	76
Chapter 8	Inferences Based on a Single Sample: Tests of Hypotheses	86
Chapter 9	Inferences Based on Two Samples: Estimation and Tests of Hypotheses	100
Chapter 10	Analysis of Variance: Comparing More Than Two Means	116
Chapter 11	Nonparametric Statistics	135
Chapter 12	The Chi-Square Test and the Analysis of Contingency Tables	150
Chapter 13	Simple Linear Regression	158
Chapter 14	Multiple Regression	180
Chapter 15	Model Building	199
ANSWERS TO SELECTED EXERCISES		219

Preface

This study guide is designed to accompany the textbook *Statistics*, by McClave and Dietrich (Sixth Edition, Dellen Publishing Company, 1994). It is intended for use as a **supplement to, not a replacement for,** the textbook. Thus, it is expected that the student will read the expository material in the textbook before referring to the corresponding sections of the study guide for reinforcement.

The following items are included for all chapters, which are titled and ordered as in the textbook:

(i) A brief **summary** highlights the concepts and terms which were introduced and explained in the textbook material.

(ii) Section-by-section **examples** provide the student with relevant applications of the statistical concepts; detailed **solutions** are given.

(iii) **Exercises** allow the student to check his or her mastery of the material in each section; **answers** for most exercises are included at the end of the study guide.

Susan L. Reiland

CHAPTER ONE

What Is Statistics?

Summary

Every statistical problem is composed of five elements: a population, a variable, a sample, an inference, and a measure of the reliability of the inference. It is the primary objective of statistics to use the information available in a **sample** to make an **inference** (that is, a decision, prediction, estimate, or generalization) about the **population** from which the sample was selected. With each inference is associated a measure of its **reliability**; each estimate or prediction will be accompanied by a bound on the prediction error, and each decision by a statement reflecting our confidence in the decision.

EXAMPLE 1.1 The New York Stock Exchange (NYSE) consists of a listing of approximately 1,500 companies which offer shares of company stock to the public. Stock brokers are interested not only in the individual stocks, but also in general trends established by the market as a whole. They often base inferences upon the daily closing prices of the group of 30 NYSE stocks that comprise the Dow Jones Industrial Index.

  a. Identify the population, the variable of interest, and the sample in the context of this problem.

  b. What are some inferences of possible interest to a stock broker? How would the reliability of the inferences be assessed?

Solution a. The **population** consists of the set of all 1,500 companies on the New York Stock Exchange. The **variable of interest** is the daily closing price of each stock. The **sample** is the set of 30 companies that comprise the Dow Jones Industrial Index. Note that the sample is a subset of the population about which inferences are to be made.

  b. The broker may be interested in estimating the average closing price of a share of stock on the NYSE, or the percentage of stocks whose closing prices were less than $25 per share on a particular day. Each estimate would be accompanied by a bound on the prediction error, that is, by a numerical value that the error of prediction is unlikely to exceed.

EXAMPLE 1.2 At ticket gates at many airports across the country is posted the following notice: "Due to deliberate overbooking of flights, there may not be a seat available for everyone who has a ticket...." Because of the loss of revenue due to "no-shows" (those who hold a reservation but fail to appear for the flight and do not notify the airline in advance), it is a common practice among airlines to overbook certain flights intentionally.

To determine how many reservations should be taken for an Atlanta-to-Houston flight, an airline wishes to develop a reliable estimate of the percentage of no-shows for this flight.

a. Specify the population of interest in this problem.

b. How might the goal of the airline be accomplished?

Solution

a. The population (which is partially conceptual) consists of all past and future realizations of the airline's Atlanta-to-Houston flight.

b. The airline could examine some recent records pertaining to the Atlanta-to-Houston flight, and select a sample of flights. The information obtained from the sample could be used to develop an estimate of the corresponding population percentage. (The details will be treated formally in subsequent chapters.)

EXAMPLE 1.3 The Environmental Protection Agency (EPA) performs gasoline mileage tests on new automobiles. In one recent test, the EPA reported that results of testing on 20 new automobiles of a particular model indicated an average mileage per gallon rating of 26.3.

a. Specify the sample, the population, and the variable of interest.

b. How might a potential automobile buyer use the information provided by the EPA?

Solution

a. The sample consists of the 20 automobiles used in this test; the population is all automobiles (of this model) manufactured. The variable of interest is the miles per gallon rating of each automobile of this particular model.

b. The buyer would be interested in predicting the average gasoline mileage that would be achieved by the particular automobile he purchases.

EXERCISES

1.1 Refer to Example 1.1. What is the **largest** possible sample of NYSE stocks that could be selected on a given day? What would you say about the reliability of inferences made from this particular sample?

1.2 A recent survey concluded that among all 1993 college graduates with degrees in the social sciences, psychology majors had the highest average starting salary, at $2050 per month.

    a. Identify the population about which the inference was made.

    b. What is needed to complete the inferential statement?

1.3 Examine some recent issues of a magazine or journal related to your field of interest. Look for applications of statistics in articles or advertisements in which decisions, estimates, or predictions are being made. Are the population and sample clearly identified? Do the inferences always contain an associated measure of reliability?

CHAPTER TWO

Methods for Describing Sets of Data

Summary

Since the objective of statistics is to use sample data to make inferences about a population, we must be able to summarize and describe the sample measurements. This chapter discussed the construction of **stem-and-leaf display** and a **histogram** to convey a visual description of a **quantitative** data set. However, because of the difficulty in obtaining measures of reliability for inferences made from graphical summaries, **numerical methods** of data description were also developed.

The following numerical measures for describing the **central tendency** of a data set were presented: the **mean**, the **median**, and the **mode**. The relative locations of the mean and median provide information about the **skewness** of a frequency distribution; the sample mean is the most preferred measure for making inferences about the central tendency of the population.

The numerical description of a set of data also requires a measure of the **variability**, or spread of the measurements about their center. The simplest measure of the variability of a quantitative data set is the **range**. However, the **variance** and the **standard deviation** are the preferred measures for purposes of making inferences about a population.

We may use the mean and standard deviation to make statements about the fraction of measurements within a specified interval. Chebyshev's rule applies to any sample of measurements, regardless of the shape of the frequency distribution. The Empirical Rule is applicable only when the frequency distribution for the sample is mound-shaped.

To describe the location of a specific measurement relative to the rest of the data set, a measure of **relative standing**, such as a **percentile** or **quartile** or **z-score**, is used. A **box plot** may be used to detect **outliers** in a data set.

The **rare event** approach to inference-making was illustrated in this chapter. The principle of this method is the following: The more unlikely it is that a particular sample came from a hypothesized population, the more strongly we tend to believe that the hypothesized population is not the one from which the sample was selected.

2.1 Types of Data

EXAMPLE 2.1 Consumer preference studies are often conducted by specialists in market research. Consider a questionnaire designed to be administered to customers at a local shopping center. The following information may be requested from each customer interviewed:

(1) Sex
(2) Age
(3) Marital status
(4) Type of store visited most frequently at the shopping center
(5) Number of visits per month customer makes to shopping center
(6) Annual income of household

a. Which of the questions will yield quantitative responses?

b. Which of the questions will yield qualitative responses?

Solution a. The responses to questions (2), (5), and (6) are ratio type data, and hence, are quantitative data.

b. The responses to questions (1), (3), and (4) are nominal type data, and thus yield qualitative data. Possible response categories for each question are suggested below.

(1) Male, Female
(3) Married, Divorced, Widowed, Single
(4) Grocery, Hardware, Department, Restaurant, Pharmacy

EXAMPLE 2.2 Classify the following examples of data as either nominal, ordinal, interval or ratio:

a. The amount paid in federal taxes last year by each of the country's 500 largest industrial corporations.

b. The highest educational degree attained by each employee of a scientific research corporation.

c. The largest selling brand of soft contact lenses in each of the states where advertising by optometrists is permitted by law.

d. The cost of a routine examination by each of a sample of 40 dentists engaged in private practice in the Northeast.

e. The amount of rainfall in Seattle, Washington, for each day during the summer months last year.

f. The degree of marital satisfaction (very happy, moderately happy, moderately unhappy, very unhappy) expressed by each married member of the faculty at a large state university.

Solution   a. Ratio   b. Ordinal   c. Nominal

  d. Ratio   e. Ratio   f. Ordinal

EXERCISES

2.1   When one applies for a major credit card, standard biographical information is requested. Classify the responses to the following items as nominal, ordinal, interval, or ratio.

   a. City of residence

   b. Length of time at current residence

   c. Bank reference

   d. Type of job

   e. Monthly salary

   f. Number of dependents

   g. Maximum amount of credit for which application is being made

2.2   Classify the following examples of data as either nominal, ordinal, interval, or ratio:

   a. The day of the week indicated by each of a sample of 200 newspaper publishers as the day for which most advertising is sold.

   b. The brand of the best-selling lawn mower at each of the 10 largest discount store chains in the Southeast.

   c. The percent increase in arrests for traffic violations by the Highway Patrol departments of the 48 contiguous United States after enactment of the 55 miles per hour speed limit.

   d. The time required for post-operative pain to be relieved in 29 surgical patients after administration of an analgesic.

2.3   Give some examples of quantitative and qualitative data that may arise from a survey in a field of particular interest to you.

2.2 Graphic Methods for Describing Quantitative Data: Histograms and Stem-and-Leaf Displays

EXAMPLE 2.3 The following data set is a sample of starting salaries (in hundreds of dollars) for recent journalism graduates at a large state university:

153	198	179	248	148	181	253	258	203
180	209	181	204	216	176	169	195	132
233	195	152	127	277	169	209		

Construct a stem-and-leaf display for the starting salary data for journalism graduates given above.

Solution The first number in the data set is 153, representing a salary of $15,300. We will designate the first two digits (15) of this number as its stem; we will call the last digit (3) its leaf, as illustrated at the right:

Stem	Leaf
15	3

The stem and leaf of the number 181 are 18 and 1, respectively. Similarly, the stem and leaf of the number 209 are 20 and 9, respectively.

The first step in forming a stem-and-leaf display for this data set is to list all stem possibilities in a column starting with the smallest stem (12, corresponding to the number 127) and ending with the largest (27, corresponding to the number 277), as shown in the figure at right. The next step is to place the leaf of each number in the data set in a row of the display corresponding to the number's stem. For example, for the number 153, the leaf 3 is placed in the stem row 15. Similarly, for the number 181, the leaf 1 is placed in the stem row 18. After the leaves of the twenty-five numbers are placed in the appropriate stem rows, the completed stem-and-leaf display will appear as shown in the figure. You can see that the stem-and-leaf display partitions the data set into sixteen categories corresponding to the sixteen stems.

Stems	Leaves
12	7
13	2
14	8
15	32
16	99
17	96
18	110
19	855
20	9439
21	6
22	
23	3
24	8
25	38
26	
27	7

EXAMPLE 2.4 The following data represent the distances (in miles) between the campus and parents' permanent residences for 50 students of a local community college:

31.4	23.6	11.8	24.9	83.8	40.0	16.6	10.4	4.9	19.0
77.0	34.9	23.3	35.5	40.1	11.9	15.5	24.1	10.1	19.3
19.6	15.3	28.9	6.6	38.8	4.0	2.4	8.5	7.6	22.3
48.9	47.1	16.8	17.8	32.3	8.9	21.3	16.9	31.5	21.5
16.8	29.0	29.6	12.1	11.8	23.4	30.3	15.6	49.3	18.4

Construct a relative frequency histogram to summarize this quantitative data set.

Solution We will construct eight measurement classes for the distance data. The smallest measurement is 2.4, so we begin the first class at 2.35. To determine the length of each measurement class, we observe that the data span a range of $83.8 - 2.4 = 81.4$; thus, each measurement class should have a length of 81.4/8, or approximately 10.2. The measurement classes, frequencies, and relative frequencies are shown in the table below.

Measurement Class	Frequency	Relative Frequency
2.35 – 12.55	13	13/50 = .26
12.55 – 22.75	15	.30
22.75 – 32.95	12	.24
32.95 – 43.15	5	.10
43.15 – 53.35	3	.06
53.35 – 63.55	0	.00
63.55 – 73.75	0	.00
73.75 – 83.95	2	.04

The relative frequencies are plotted as rectangles over the corresponding measurement classes in the histogram at right.

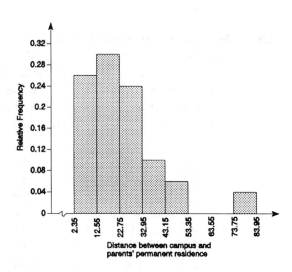

EXAMPLE 2.5 The relative frequency histogram shown at the right summarizes the distribution of 1993 taxable income for a sample of 200 research technicians employed by the federal government.

What proportion of the sampled technicians had 1993 incomes in excess of $40,000?

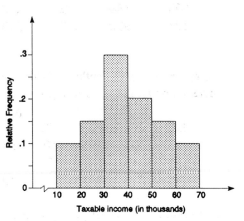

Solution We first recall that the proportion of the total area that falls above an interval in the histogram is equal to the relative frequency of measurements falling within that interval. In this example, the interval from $40,000 – $70,000 contains (.2 + .15 + .1) = .45 of the total area under the histogram. Thus, .45 of the sampled individuals had 1993 incomes within the interval $40,000 – $70,000, i.e., in excess of $40,000.

EXERCISES

2.4   Consider the following sample data:

213	228	241	268	234	303
274	316	319	320	227	226
224	267	303	266	265	237
288	291	285	270	254	215

a.   Using the first two digits of each number as a stem, list the stem possibilities in order.

b.   Place the leaf for each observation in the appropriate stem row to form a stem-and-leaf display.

2.5   Hospitals are required to file a yearly cost report in order to obtain reimbursement from the state for patient bills paid through the Medicare, Medicaid, and Blue Cross programs. Many factors contribute to the amount of reimbursement that the hospital receives. One important factor is bed size (i.e., the total number of beds available for patient use). The data below represent the bed sizes for fifty-four hospitals that were satisfied with their cost report reimbursements last year. Construct a stem-and-leaf display for the bed size data.

303	550	243	282	195	310
288	188	190	335	473	169
292	492	200	478	182	172
231	375	171	262	198	313
600	264	311	371	145	242
278	183	215	719	519	382
249	350	99	218	300	450
337	330	252	400	514	427
533	930	319	210	550	488

2.6   Forty drivers from southern California reported the following amounts (in dollars) spent on annual automobile insurance premiums:

320	297	423	193	403	203	443	179	276	223
163	208	198	278	303	199	195	185	297	297
241	236	270	287	403	190	248	253	250	323
285	236	383	238	220	347	291	410	230	263

Construct a relative frequency histogram for these data, using six measurement classes.

2.7   a.   Repeat Exercise 2.6, using four measurement classes.

       b.   Repeat Exercise 2.6, using eight measurement classes.

       c.   How do the graphical representations of the data differ as the number of measurement classes is changed?

2.3   Numerical Measures of Central Tendency

EXAMPLE 2.6   Eight employees of a university administration were asked to report the number of miles traveled using mass transportation during a typical week. The responses were as follows:

$$50, \; 0, \; 100, \; 65, \; 420, \; 70, \; 0, \; 100$$

Compute the mean for this sample of measurements.

Solution   The mean of the eight measurements is

$$\bar{x} = \frac{\sum x}{n} = \frac{50 + 0 + 100 + 65 + 420 + 70 + 0 + 100}{8} = \frac{805}{8} = 100.6$$

These employees travel, on the average, 100.6 miles per week using mass transportation.

EXAMPLE 2.7   The unemployment rates (in percent) for the six largest cities of a southern state during the fourth quarter last year were reported as follows:

$$7.8 \quad 8.3 \quad 8.8 \quad 6.9 \quad 9.2 \quad 8.5$$

Compute the mean unemployment rate for this group of cities.

Solution   The sample mean is computed as follows:

$$\bar{x} = \frac{\sum x}{n} = \frac{7.8 + 8.3 + 8.8 + 6.9 + 9.2 + 8.5}{6} = 8.25$$

Among this group of six southern cities, the average unemployment rate was 8.25% during the stated period.

EXAMPLE 2.8   Many companies have a majority of young employees who receive lower salaries than more experienced people. Thus, there is often an imbalance between the bottom and top levels of a company's salary structure.

Consider the following salaries (in thousands of dollars) for seven employees of an advertising agency:

$$28.1, \; 27.9, \; 23.9, \; 24.4, \; 65.2, \; 25.3, \; 24.8$$

a. Compute the mean salary.

b. Compute the median salary.

c. Which measure provides a more realistic measure of the central tendency of the data?

Solution   a. $\bar{x} = \dfrac{\sum x}{n} = \dfrac{28.1 + 27.9 + 23.9 + 24.4 + 65.2 + 25.3 + 24.8}{7} = 31.4$

The mean (average) salary is $31,400.

b. It is first required to arrange the seven measurements in ascending order:

23.9, 24.4, 24.8, 25.3, 27.9, 28.1, 65.2

Since the number of measurements is odd, the median is the middle observation. Thus, the median salary for this sample of employees is $25,300.

c. The calculation of the median salary was less sensitive to the extremely large observation (65.2) in the sample, and thus provides a more appropriate measure of the central tendency of the salary data.

EXAMPLE 2.9   Refer to Example 2.8. Suppose the agency hires an additional junior executive, at a salary of $31,000. Determine the median salary for the eight employees.

Solution   The new salary is recorded in thousands of dollars as 31.0. Then the sample is ordered as follows:

23.9, 24.4, 24.8, 25.3, 27.9, 28.1, 31.0, 65.2

The number of measurements is even, so the median is the average of the middle two observations:

$$\text{median} = \dfrac{25.3 + 27.9}{2} = 26.6$$

The median salary is increased to $26,600 with the hiring of the new employee.

EXAMPLE 2.10   The following data show the number of hospitalizations required during the past two years by the eight supervisory air traffic controllers at an international airport:

0, 3, 0, 0, 1, 0, 2, 1

Calculate the mode for these data.

Solution   The measurement 0 is observed four times, more frequently than any other value. Thus, the mode for this data set is 0.

Methods for Describing Sets of Data

EXAMPLE 2.11 The following is the relative frequency distribution of 500 Civil Service examination scores achieved by recent job applicants with the United States Postal Service. Determine the modal examination score for this group of applicants.

Class	Relative Frequency
19.5 – 39.5	.12
39.5 – 59.5	.28
59.5 – 79.5	.36
79.5 – 99.5	.24

Solution The modal class, the interval containing the most measurements, is 59.5 – 79.5. Thus, the mode is equal to the midpoint of this class interval, or (59.5 + 79.5)/2 = 69.5.

EXERCISES

2.8 The following data represent the daily profits on 15 randomly chosen business days for a boutique specializing in leather gift items:

$370.11 $380.37 $351.35 $278.12 $269.96
 371.29   388.37   461.73   369.25   371.46
 381.00   464.37   279.47   281.70   278.83

Compute the mean profit for these 15 business days.

2.9 The number of hours of "down-time" for a university computing system was recorded for six consecutive days, during which time new tape drive equipment was being installed and tested. Compute the average "down-time" for the following results:

10.2, 6.3, 4.0, 2.1, 8.5, 4.6

2.10 The following data represent the number of days beyond the estimated completion date required to finish six building construction projects. (Negative values indicate the construction was complete before the target date.)

4, 23, −6, 12, −20, 31

Compute the median for this sample.

2.11 The numbers of automobile insurance claims received at a district office were recorded for 10 consecutive business days as follows:

0, 2, 3, 1, 3, 2, 4, 3, 2, 3

Compute the mode for this set of data.

2.12 Give an example of a situation in which the mode would be a more appropriate measure of central tendency than the mean.

2.13 Suppose you are interviewed for a job and are offered a starting salary of $17,500 per year. What would be your reaction to this offer if you had the additional information that:

   a. The mean starting salary for all jobs of this type is $19,000.

   b. The modal starting salary is $17,500.

   c. The median starting salary is $17,000.

   d. The mean starting salary is $19,000, the median is $17,000, and the mode is $16,000.

   e. The mean starting salary is $17,000 and the median is $18,000.

2.4 Numerical Measures of Variability

EXAMPLE 2.12 The following histograms show the relative frequency of the numbers of chapters in introductory statistics textbooks that were published by two publishers over the past decade. Compute the range in numbers of chapters for the two publishers and comment on why the range is often an inadequate measure of the variability of a data set.

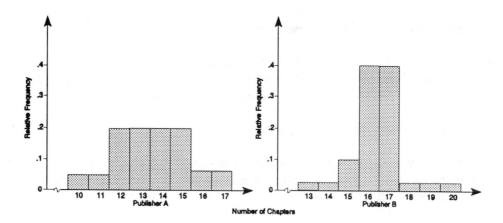

Solution   For Publisher A,

   range = largest observation − smallest observation
         = 17 − 10
         = 7 chapters

   For Publisher B,

   range = 20 − 13
         = 7 chapters

Methods for Describing Sets of Data

Although the ranges of the two data sets are equal, the measurements for Publisher A are more spread out than are those for Publisher B. A more informative numerical measure of variability is necessary for these data sets.

EXAMPLE 2.13 The following measurements represent the number of hours of petty leave taken during the past month by seven employees of a large department store:

5, 8, 0, 3, 10, 6, 3

Calculate the sample variance and standard deviation for this data set.

Solution It is necessary first to compute the sample mean number of hours of leave taken by the employees:

$$\bar{x} = \frac{\sum x}{n} = \frac{5 + 8 + 0 + 3 + 10 + 6 + 3}{7} = \frac{35}{7} = 5$$

The calculations required for the sample variance, s^2, are shown in the following table:

Data x	$x - \bar{x}$	$(x - \bar{x})^2$
5	5 - 5 = 0	$(0)^2$ = 0
8	8 - 5 = 3	$(3)^2$ = 9
0	0 - 5 = -5	$(-5)^2$ = 25
3	3 - 5 = -2	$(-2)^2$ = 4
10	10 - 5 = 5	$(5)^2$ = 25
6	6 - 5 = 1	$(1)^2$ = 1
3	3 - 5 = -2	$(-2)^2$ = 4
$\sum x = 35$		$\sum (x - \bar{x})^2 = 68$

Now, the sample variance is

$$s^2 = \frac{\sum (x - \bar{x})^2}{n - 1} = \frac{68}{7 - 1} = \frac{68}{6} = 11.33$$

and the sample standard deviation is given by

$$s = \sqrt{s^2} = \sqrt{11.33} = 3.37$$

EXAMPLE 2.14 For a particular data set, the following information is known:

$$n = 6, \quad \sum x^2 = 176, \quad \sum x = 18$$

Compute the sample variance and standard deviation.

Solution The shortcut formula for sample variance will be applied:

$$s^2 = \frac{\sum x^2 - \frac{(\sum x)^2}{n}}{n-1} = \frac{176 - \frac{(18)^2}{6}}{6-1} = \frac{176 - \frac{324}{6}}{5} = \frac{174 - 54}{5} = \frac{122}{5} = 24.4$$

Thus, the sample variance is 24.4 and the sample standard deviation is

$$s = \sqrt{s^2} = \sqrt{24.4} = 4.94$$

EXAMPLE 2.15 Five senior employees of an aerospace corporation received the following percentage increases in their salaries during the last fiscal year:

14, 9, 16, 13, 15

Use the shortcut formula to compute s^2 for this data set.

Solution For these $n = 5$ measurements,

$$\sum x = 14 + 9 + 16 + 13 + 15 = 67$$

and

$$\sum x^2 = (14)^2 + (9)^2 + (16)^2 + (13)^2 + (15)^2 = 927$$

Then

$$s^2 = \frac{\sum x^2 - \frac{(\sum x)^2}{n}}{n-1} = \frac{927 - \frac{(67)^2}{5}}{4} = \frac{927 - 897.8}{4} = 7.3$$

EXERCISES

2.14 The Chamber of Commerce of a large city is interested in the monthly rental rates of three-bedroom, two-bath apartments in the area. Managers of seven apartment complexes submitted the following rental rates of their three-bedroom, two-bath units:

$400, $465, $335, $305, $560, $410, $440

Compute the range in rental rates for this sample.

2.15 Records for eight multinational corporations indicated the following percentages of total company assets held in the United States:

80   46   76   54   55   44   58   51

Compute the variance and standard deviation for this sample of measurements.

2.16 Given the following information about a data set of interest, calculate the sample variance and standard deviation:

$n = 12$, $\sum x = 435$, $\sum x^2 = 19{,}840$

2.17 Drivers for an interstate trucking company reported the following numbers of miles driven during the past 24 hours:

420, 360, 380, 290, 450, 360

Compute the sample variance and standard deviation of the mileages, using the shortcut formula.

2.5 Interpreting the Standard Deviation

EXAMPLE 2.16 Recent real estate figures have shown that the average length of residency in a given home is 13 years, and the standard deviation is 4 years. If we make no assumptions about the frequency distribution of lengths of residency, what can be said about the fraction of homeowners who live in their houses between 9 and 17 years? Between 5 and 21 years? Between 1 and 25 years?

Solution To apply the aids for interpreting the value of a standard deviation, we first form the intervals

$$(\bar{x} - s, \bar{x} + s) = (13 - 4, 13 + 4) = (9, 17)$$

$$(\bar{x} - 2s, \bar{x} + 2s) = (13 - 8, 13 + 8) = (5, 21),$$

and

$$(\bar{x} - 3s, \bar{x} + 3s) = (13 - 12, 13 + 12) = (1, 25)$$

According to Chebyshev's rule, we may make the following statements:

a. It is possible that very few of the measurements will fall within the one standard deviation interval about the mean (9, 17).
b. At least 3/4 of the measurements fall within (5, 21), the two standard deviation interval about the mean. In terms of the problem, we conclude that at least 3/4 of the homeowners live in their houses between 5 and 21 years.
c. At least 8/9 of the homeowners reside in their homes between 1 and 25 years.

EXAMPLE 2.17 Suppose that a frequency histogram of the lengths of home residency in a particular region of the country shows the distribution to be mound-shaped, with mean 13 years and standard deviation 4 years. What fraction of the lengths of residency would be expected to fall within each of the intervals specified in Example 2.16? Compare your results with those obtained in the previous example.

Solution We may use the Empirical Rule, which applies to samples with mound-shaped frequency distributions.

a. Approximately 68% of the homeowners reside in their homes between 9 and 17 years; i.e., approximately 68% of the measurements will fall within one standard deviation of the mean.
b. Approximately 95% of the homeowners live in their homes between 5 and 21 years.

c. Essentially all the homeowners in this region live in their homes between 1 and 25 years.

When nothing is known about the shape of the frequency distribution, Chebyshev's rule regarding the fraction of observations that will lie within the one standard deviation interval about the mean is not very informative (Example 2.16a). However, if there is evidence that the frequency distribution is mound-shaped, we may make the more meaningful statement that approximately 68% of the measurements will fall within one standard deviation of the mean. Similarly, Chebyshev's rule guarantees only that at least 75% of the measurements will be within two standard deviations of the mean; if the distribution is mound-shaped, we have the stronger conclusion that approximately 95% of the measurements will lie within this interval.

EXAMPLE 2.18 In a test designed to assess small motor coordination, kindergarten pupils are timed on their performance of a specific task. It is known that the time required to complete the task has an approximate mound-shaped distribution with mean 36 seconds and standard deviation 6 seconds. If a child requires more than 48 seconds to perform the task, he or she is recommended for further developmental testing. Approximately what percent of kindergarten pupils fail to accomplish the task within 48 seconds?

Solution The distribution of time required by kindergarten pupils to perform the task may be visualized as shown at the right.

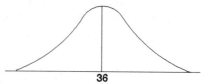

The distribution is symmetric, with the mean located at the center of the distribution.

Now we apply the Empirical Rule to conclude that approximately 95% of the pupils' task performance times fall within the interval $(\bar{x} - 2s, \bar{x} + 2s) = (36 - 12, 36 + 12) = (24, 48)$ seconds:

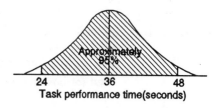

We observe that approximately 5% of the distribution remains to be divided between the two tails. The symmetry of the distribution implies that half of this amount, or approximately 2.5%, will lie below 24 seconds and approximately 2.5% will lie above 48 seconds.

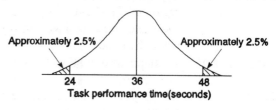

Thus, approximately 2.5% of the pupils will require more than 48 seconds to complete the task, and will be recommended for further testing.

Methods for Describing Sets of Data

EXERCISES

2.18 Refer to Example 2.18. How long should the test administrator allow for the performance of the task if it is desired that approximately 84% of the kindergarten pupils complete the task?

2.19 First-year sales for introductory mathematics textbooks have had an average volume of 1,200 books and a standard deviation of 500 books.

    a. What can be said about the fraction of mathematics texts that have first-year sales between 700 and 1,700 books? Between 200 and 2,200 books?

    b. Suppose that the distribution of first-year sales for mathematics texts is known to be mound-shaped. What percentage of mathematics texts would you expect to have first-year sales between 700 and 1,700 books? Between 200 and 2,200 books?

    c. Assume the frequency distribution of first-year sales is mound-shaped. What fraction of mathematics texts would you expect to have first-year sales in excess of 2,200 books? Less than 700 books?

2.20 The chairman of the Greater Miami Chamber of Commerce asked 500 attendees at a political convention to keep records of all expenditures made while in attendance at the three-day convention. A relative frequency histogram of the results showed the distribution of expenditures, which ranged between $200 and $750, to be approximately mound-shaped. Use this information to estimate the mean and standard deviation of the distribution of expenditures for this sample.

2.6 Measures of Relative Standing

EXAMPLE 2.19 A local bank advertises that, among all banking institutions in the country, it is in the 77th percentile of the distribution of amount of capital used for consumer (non-mortgage) loans. Explain how this locates the bank in the distribution of capital used for consumer loans.

Solution The interpretation is that 77% of all banking institutions use less capital for consumer loans than does the local bank, while $(100 - 77) = 23\%$ of all banks use more capital for this purpose.

EXAMPLE 2.20 A sample of 200 homes listed for sale in a particular area showed an average selling price of $67,000 and a standard deviation of $9,000. Suppose your house is selling for $58,000 and the house you wish to purchase is selling for $94,000. Compute the z-scores for these two sale prices.

Solution For this sample of measurements, we have $\bar{x} = 67{,}000$ and $s = 9{,}000$. Thus, for your house, $x = 58{,}000$ and

$$z = \frac{x - \bar{x}}{s} = \frac{58{,}000 - 67{,}000}{9{,}000} = -1$$

The price of your house is one standard deviation **below** the mean selling price for this sample.

For the home you wish to purchase, $x = 94{,}000$ and

$$z = \frac{94{,}000 - 67{,}000}{9{,}000} = 3$$

This home is priced 3 standard deviations **above** the mean selling price for the sample.

EXAMPLE 2.21 In Example 2.18, it was stated that the time required by kindergarten pupils to complete a specific task has a mound-shaped distribution with mean 36 seconds and standard deviation 6 seconds. Suppose a particular kindergarten pupil's z-score for the time required was 1.5. How long did he or she required to complete the task?

Solution Let x denote the time required by the kindergarten pupil to finish the task. The z-score corresponding to this value of x is known to be 1.5. Now, since the mean and standard deviation of the distribution are given, it is required to solve the following equation for x:

$$z = \frac{x - \bar{x}}{s} \text{ or } 1.5 = \frac{x - 36}{6}$$

Elementary algebra yields the solution

$$x = 6(1.5) + 36 = 45$$

Thus, the pupil completed the task in 45 seconds.

EXAMPLE 2.22 An oil additive is tested by the Department of Transportation for its effect on gasoline mileage. Test results indicated that the increases in miles per gallon (mpg) obtained with the additive follow an approximate mound-shaped distribution with mean 2.2 mpg and standard deviation .5 mpg. A local service station attendant claims the additive will increase mileage by 4 mpg for most automobiles. Is this a reasonable claim?

Solution Although the increased mileage predicted by the attendant at first looks very attractive, let us consider the relative standing of this prediction in the distribution of increases in mpg by Department of Transportation testing. The predicted increase of $x = 4$ mpg has z-score

$$z = \frac{4 - 2.2}{.5} = 3.6$$

Now we know that for mound-shaped distributions essentially **all** the measurements lie within 3 standard deviations of the mean; i.e., almost all measurements have z-scores between -3 and 3. An increase of 4 mpg lies 3.6 standard deviations above the mean and would thus be a very **rare** event. A consumer would be advised to scrutinize carefully the attendant's claim.

Methods for Describing Sets of Data

EXAMPLE 2.23 Refer to Example 2.3. Find the lower quartile, median, and upper quartile for the twenty-five starting salaries.

Solution We first rank the twenty-five observations from smallest to largest:

127	132	148	152	153
169	169	176	179	180
181	181	195	195	198
203	204	209	209	216
233	248	253	258	277

To find Q_L, calculate $(1/4)(n + 1) = (1/4)(25 + 1) = 6.5$, and round to the integer 7. [Since $(1/4)(n + 1)$ is an integer plus 1/2 in this case, we round upward.] Thus, the lower quartile is the observation with rank 7 in the data set: $Q_L = 169$.

The median is the middle (i.e., the thirteenth) ranked observation: median = 195.

To find Q_U, calculate $(3/4)(n + 1) = 19.5$. Since this quantity is equal to an integer plus 1/2, we round downward and obtain the upper quartile as the nineteenth ranked observation: $Q_U = 209$.

EXERCISES

2.21 a. What is the z-score corresponding to the 97.5th percentile of a mound-shaped frequency distribution?

b. What is the z-score corresponding to the 16th percentile of a mound-shaped frequency distribution?

c. A z-score of 0 corresponds to which percentile of a mound-shaped distribution?

2.22 Refer to Example 2.18. Compute the z-scores associated with the following task performance times:

a. 48 seconds   b. 30 seconds   c. 36 seconds

d. 32 seconds   e. 1 minute

2.23 The distribution of sugar content for cereals marketed by a particular food processor is known to have a mean of 18% and a standard deviation of 1.6%.

a. What is the z-score corresponding to a sugar content of 17.2%?

b. Suppose a particular cereal has a sugar content with a z-score of $-.4$. To what sugar content does this z-score correspond?

2.24 During the past six years, return rates on government-associated investments have followed a mound-shaped distribution with mean 8.8% and standard deviation 1.4%. Suppose a broker is trying to interest you in a government investment for which he predicts a return rate of 14.4%. Why might you be skeptical of his prediction?

2.25 Refer to Exercise 2.4. Determine the lower quartile, median, and upper quartile for the data set.

2.26 Refer to the bed size data given in Exercise 2.5. Find the lower quartile, median, and upper quartile for the fifty-four observations.

2.7 Box Plots: Graphic Descriptions Based on Quartiles (Optional)

EXAMPLE 2.24 Refer to the starting salary data for journalism majors given in Example 2.3. Construct a box plot for the data.

Solution The first step is to construct a box with Q_L and Q_U located at the lower corners. In Example 2.23, we found Q_L and Q_U to be 169 and 209, respectively. The interquartile range is then

$$IQR = 209 - 169 = 40$$

The second step is to locate the inner fences, which lie a distance of 1.5(IQR) = 1.5(40) = 60 below Q_L and above Q_U. These values, $Q_L - 1.5(IQR) = 169 - 60 = 109$ and $Q_U + 1.5(IQR) = 209 + 60 = 269$, are located on the box plot.

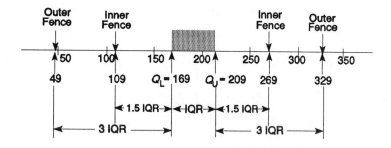

The third step is to locate the outer fences, which lie a distance of 1.5(IQR) = 60 below the lower inner fence and above the upper inner fence. Thus, the outer fences for this data set are located at 49 and 329, as indicated in the figure.

EXERCISES

2.27 Refer to Exercises 2.4 and 2.25.

    a. Compute the interquartile range for the data set.

    b. Construct a box plot for the data.

2.28 Refer to Exercises 2.5 and 2.26.

    a. Compute the interquartile range for the bed size data.

    b. Construct a box plot for the data.

CHAPTER THREE

Probability

Summary

This chapter presented the basic concepts and tools of **probability**, which will allow us to make inferences about the population from an observed sample. In addition, the theory of probability will often be used in measuring the reliability of inferences.

The notions of **experiments, events, event relations,** and **random sampling** were defined. Rules for assigning probabilities to events in the sample space and for computing the probability of an event of interest were presented. Four counting rules for determining the number of simple events in an experiment were developed.

3.1 Events, Sample Spaces, and Probability

EXAMPLE 3.1 A major department store chain is planning to open a store in a new city. Five cities are being considered: Boston, Atlanta, Dallas, Cleveland, and Los Angeles.

    a.   List the simple events associated with this experiment.

    b.   Assign a probability to each simple event, assuming each city has an equal chance of being selected.

    c.   Compute the probability of each of the following events:

        G: {Dallas is chosen}
        H: {A southern city is chosen}
        K: {Los Angeles is not chosen}

Solution   a.   The experiment consists of choosing one city at random from the five specified. The simple events may be enumerated as follows:

        (1)   Boston is chosen.
        (2)   Atlanta is chosen.
        (3)   Dallas is chosen.
        (4)   Cleveland is chosen.
        (5)   Los Angeles is chosen.

    b.   Since there are five equally likely simple events, each must be assigned a probability of 1/5; i.e., P(Boston) = P(Atlanta) = P(Dallas) = P(Cleveland) = P(Los Angeles) = 1/5. Note that the two requirements for assigning probabilities to simple events are satisfied: each probability is between 0 and 1, and the probabilities of all the simple events sum to 1.

c. $P(G) = P(\text{Dallas}) = 1/5$

$P(H) = P(\text{Atlanta}) + P(\text{Dallas}) = 1/5 + 1/5 = 2/5$

$P(K) = P(\text{Boston}) + P(\text{Atlanta}) + P(\text{Dallas}) + P(\text{Cleveland})$
$= 1/5 + 1/5 + 1/5 + 1/5 = 4/5$

EXAMPLE 3.2 The city council of a particular community consists of five elected residents of the community, two of whom are land developers. The city mayor plans to select two members at random from the council to study and make recommendations on land use rezoning requests. The composition of this subcommittee is of particular interest.

a. List the simple events in the sample space for this experiment.

b. Assuming that each pair of city council members has an equal chance of being selected, assign probabilities to each simple event.

c. Compute the probabilities of the following events of interest:

D: {Both land developers are selected}
F: {At least one land developer is selected}
G: {No land developer is selected}

Solution a. We will label the city council members C_1, C_2, C_3, L_1, and L_2, where the two land developers are represented by L_1 and L_2. The experiment consists of randomly selecting two council members; the ten simple events may be represented as unordered pairs of members:

$(C_1 C_2)$ $(C_1 C_3)$ $(C_1 L_1)$ $(C_1 L_2)$ $(C_2 C_3)$
$(C_2 L_1)$ $(C_2 L_2)$ $(C_3 L_1)$ $(C_3 L_2)$ $(L_1 L_2)$

b. The ten simple events are assumed equally likely; thus, each is assigned a probability of 1/10.

c. $P(D) = P(L_1 L_2) = 1/10$

$P(F) = P(C_1 L_1) + P(C_1 L_2) + P(C_2 L_1) + P(C_2 L_2) + P(C_3 L_1)$
$\quad\quad + P(C_3 L_2) + P(L_1 L_2)$
$= 7/10$

$P(G) = P(C_1 C_2) + P(C_1 C_3) + P(C_2 C_3) = 3/10$

EXAMPLE 3.3 A retail grocer has decided to market organic "health foods" and will purchase a new line of products from each of two suppliers. Unknown to the grocer, the two suppliers are in financial distress. Past experience has shown that, for firms with similar credit histories, the probability that bankruptcy proceedings will be initiated within one year is .7. We are interested in observing the financial progress of the two suppliers over the next year.

For this experiment, the simple events and their associated probabilities are as follows (B_1: Supplier 1 declares bankruptcy; N_1: Supplier 1 does not declare bankruptcy, etc.):

Simple Events	Probabilities
(B_1, B_2)	.49
(B_1, N_2)	.21
(N_1, B_2)	.21
(N_1, N_2)	.09

Compute the probabilities of each of the following events:

D: {Neither supplier declares bankruptcy during the next year}
F: {At least one supplier declares bankruptcy during the next year}

Solution

$P(D) = P(N_1, N_2) = .09$

$P(F) = P(B_1, B_2) + P(B_1, N_2) + P(N_1, B_2) = .49 + .21 + .21 = .91$

EXAMPLE 3.4 Epidemiologists at a medical center in the Northeast are interested in the etiology of anthracosis ("black lung" disease), caused by the inhalation of coal dust. All the coal miners in a particular community were screened and classified according to their length of employment in the coal mines and whether symptoms of anthracosis were present. The results are shown in the following table:

		Length of Employment	
		Less than 3 Years	3 Years or More
Anthracosis Symptoms	Present	18%	52%
	Absent	24%	6%

Suppose a coal miner is selected at random from this community, and his length of employment and presence or absence of symptoms are noted.

a. List the simple events for this experiment.

b. Let B be the event that the worker has symptoms of anthracosis. Find $P(B)$.

c. Let C be the event that the worker has spent at least 3 years in the mines. Find $P(C)$.

d. Let D be the event that the worker has spent less than 3 years in the mines and has no symptoms of anthracosis. Find $P(D)$.

Solution

a. The four simple events for this experiment may be written as follows:

$$(L, P), (L, A), (M, P), (M, A)$$

where (L, P) indicates that the worker has been employed **less** than 3 years and symptoms of anthracosis are **present**; (M, A) indicates the worker has been employed 3 years or **more** and symptoms of anthracosis are **absent**, etc.

b. $P(B) = P(L, P) + P(M, P) = .18 + .52 = .70$

c. $P(C) = P(M, P) + P(M, A) = .52 + .06 = .58$

d. $P(D) = P(L, A) = .24$

EXERCISES

3.1 Refer to Example 3.1. Suppose the chain will open new stores in two of the candidate cities and that the selection of the pair of cities will be random.

    a. Define the experiment and list the simple events.

    b. Compute the probability that neither Boston nor Cleveland is chosen.

3.2 A retailer of stereo equipment has observed that the probability a particular customer will request long-term financing is .8. This afternoon the retailer will make sales to each of 3 customers and will note their preferences for financial arrangements. The simple events for this experiment, together with their associated probabilities, are as follows (F_1: Customer 1 requests long-term financing, N_1: Customer 1 does not request long-term financing, etc.):

Simple Events	Probabilities
(F_1, F_2, F_3)	.512
(F_1, F_2, N_3)	.128
(F_1, N_2, F_3)	.128
(F_1, N_2, N_3)	.032
(N_1, F_2, F_3)	.128
(N_1, F_2, N_3)	.032
(N_1, N_2, F_3)	.032
(N_1, N_2, N_3)	.008

Calculate the probabilities of the following events of interest:

    A: {At least one customer requests long-term financing}
    B: {Exactly one customer requests long-term financing}
    C: {No customer requests long-term financing}

3.2 Unions and Intersections

EXAMPLE 3.5 The following list of schools in the Atlantic Coast Conference shows the state in which each school is located and the brand of basketball shoe worn by the men's varsity basketball team.

School	State	Shoe Brand
Maryland	Maryland	Adidas
Virginia	Virginia	Converse
North Carolina	North Carolina	Nike
North Carolina State	North Carolina	Converse
Wake Forest	North Carolina	Adidas
Duke	North Carolina	Nike
Clemson	South Carolina	Adidas
Georgia Tech	Georgia	Converse
Florida State	Florida	Nike

One school is selected at random and the following are defined:

A: {The school is located in North Carolina}
B: {The school's men's basketball team wears Adidas}
C: {The school's men's basketball team wears Converse}

Describe the characteristics implied by the following compound events:

a. $A \cup B$
b. $A \cap B$
c. $A \cup B \cup C$
d. $A \cap C$

Solution

a. $A \cup B$: {Either the school is located in North Carolina, or the men's basketball team wears Adidas, or both}

Therefore, the schools in $A \cup B$ are Maryland, North Carolina, North Carolina State, Wake Forest, Duke, and Clemson.

b. $A \cap B$: {The school is located in North Carolina and the men's basketball team wears Adidas}

Therefore, the only school in $A \cap B$ is Wake Forest.

c. $A \cup B \cup C$: {The school is located in North Carolina, or the men's basketball team wears Adidas, or the men's basketball team wears Converse}

Therefore, the schools in $A \cup B \cup C$ are Maryland, Virginia, North Carolina, North Carolina State, Duke, Wake Forest, Clemson, and Georgia Tech.

d. $A \cap C$: {The school is located in North Carolina and the men's basketball team wears Converse}

Therefore, the only school in $A \cap C$ is North Carolina State.

EXAMPLE 3.6 Refer to Example 3.5. If each school is equally likely to be chosen, find:

a. $P(A \cup B)$ b. $P(A \cap C)$

Solution a. Since the probability of an event is the sum of the probabilities of the simple events of which the event is composed,

$$P(A \cup B) = P(\text{Maryland}) + P(\text{North Carolina})$$
$$+ P(\text{North Carolina State}) + P(\text{Wake Forest})$$
$$+ P(\text{Duke}) + P(\text{Clemson})$$
$$= 1/9 + 1/9 + 1/9 + 1/9 + 1/9 + 1/9$$
$$= 6/9 = 2/3$$

b. $P(A \cap C) = P(\text{North Carolina State}) = 1/9$

EXAMPLE 3.7 The percentages of undergraduate students enrolled in 4-year colleges by age and full-time/part-time status are shown in the table.

		Status	
		Full-time	Part-time
	18 – 19	27.4	.9
Age	20 – 21	26.9	1.8
	<18 or >21	17.5	25.5

Define the following events:

   A: {A student attends college full-time}
   B: {A student's age is less than 18 or 20 or more}

a. Find $P(A)$ and $P(B)$.

b. Find $P(A \cup B)$.

c. Find $P(A \cap B)$.

Solution   The objective is to characterize the status and age of undergraduates at 4-year colleges. To accomplish this we define the experiment to consist of selecting an undergraduate from the collection of all undergraduates at 4-year colleges and observing which status and age class he or she occupies. The simple events are the six different age-status combinations:

E_1: {18–19 yrs, full-time}
E_2: {20–21 yrs, full-time}
⋮   ⋮
E_6: {<18 or >21, part-time}

You may want to verify that the simple event probabilities sum to 1.

To assign probabilities to the simple events, we note that if we randomly choose an undergraduate, the probability that he or she will occupy a particular age-status classification is the relative frequency of undergraduates in the classification. These relative frequencies are given in the table. Thus,

$P(E_1)$ = Relative frequency of undergraduates in age-status class {18–19 yrs, full-time} = .274
$P(E_2)$ = .269

and so forth.

a. To find $P(A)$, we first find the simple events that comprise event A. Since A is defined as {A student attends college full-time}, we observe that A contains the three simple events in the first column of the table. The probability of A is the sum of the probabilities of the simple events in A:

$P(A) = .274 + .269 + .175 = .718$

Similarly, event B consists of the simple events in the second and third rows of the table:

$P(B) = .269 + .018 + .175 + .255 = .717$

b. The event $A \cup B$ consists of the simple events in A or B or both. Therefore, $A \cup B$ consists of all undergraduates whose status is full-time or whose age is less than 18 or 20 or more. In the given table, this is any simple event found in the first column or the last two rows. Therefore,

$P(A \cup B) = .274 + .269 + .175 + .018 + .255 = .991$

c. The event $A \cap B$ consists of all simple events in both A and B, that is, undergraduates whose status is full-time and whose age is less than 18 or 20 or more. In the given table, this is any simple event in the first column and the last two rows. Therefore,

$P(A \cap B) = .269 + .175 = .444$

Probability

EXERCISES

3.3 You draw one M&M candy from a large bag full of M&M's and record its color. The outcome is governed by the following table of probabilities supplied by the manufacturer of M&M's:

Color	Brown	Red	Yellow	Green	Orange	Tan
Probability	.3	.2	.2	.1	.1	.1

Find the probability that:

a. You select brown or red.

b. You select yellow, red, or tan.

c. You select brown, red, yellow, green, orange, or tan.

d. You select yellow and orange.

3.4 The type of medical care a patient receives may depend on the age of the patient. A large study of men who had preliminary indications of prostate cancer based on a new blood test investigated whether or not each man received follow-up treatment. The table lists the probabilities of follow-up treatment based on age.

Age	Follow-up Treatment	No Follow-up Treatment
Under 70	.321	.124
70 or over	.365	.190

Define the following events:

A: {A patient in this study is under 70}
B: {A patient in this study receives follow-up treatment}

a. Find $P(A)$ and $P(B)$.

b. Find $P(A \cup B)$.

c. Find $P(A \cap B)$.

3.3 The Additive Rule and Mutually Exclusive Events
3.4 Complementary Events

EXAMPLE 3.8 Suppose events A and B are such that $P(A) = 1/4$, $P(B) = 1/3$, and $P(A \cap B) = 1/8$.

    a. Compute $P(A \cup B)$.

    b. Are events A and B mutually exclusive?

Solution   a. The Additive Rule of Probability yields

$$P(A \cup B) = P(A) + P(B) - P(A \cap B) = 1/4 + 1/3 - 1/8 = 11/24$$

    b. $P(A \cap B) \neq 0$; thus, events A and B are not mutually exclusive.

EXAMPLE 3.9 One hundred members of a small suburban country club were surveyed to determine the pattern of use for its recreational facilities, which consist of a golf course and swimming pool. The following results were reported.

    92 members regularly use at least one of the facilities
    25 members regularly use the golf course
    86 members regularly use the swimming pool

How many members regularly use **both** facilities?

Solution We define the following events of interest:

    A: {Member regularly uses golf course}
    B: {Member regularly uses swimming pool}

In terms of these events, the following probabilities are known:

$$P(A) = .25, \; P(B) = .86, \; P(A \cup B) = .92$$

To determine $P(\text{Member regularly uses both facilities}) = P(A \cap B)$, we solve for $P(A \cap B)$ in the Additive Rule of Probability:

$$P(A \cup B) = P(A) + P(B) - P(A \cap B)$$

or

$$.92 = .25 + .86 - P(A \cap B)$$
$$.92 = 1.11 - P(A \cap B)$$

or

$$P(A \cap B) = .19$$

Thus, $P(A \cap B) = .19$ and 19 of the 100 members surveyed regularly use both the golf course and swimming pool.

EXAMPLE 3.10 Assembly line workers in a large factory were rated according to the adequacy of their breakfast and their work efficiency at 10 A.M. Results are shown in the following table:

		Efficiency at 10 A.M.	
		Satisfactory	Unsatisfactory
Breakfast	Adequate	28%	23%
	Inadequate	20%	29%

One factory worker is selected at random, and the following events are defined:

A: {Worker's efficiency is unsatisfactory}
B: {Worker's breakfast is adequate}

Define the following events in terms of the problem:

a. A' b. B' c. $A \cap B'$

Solution

a. A': {Worker's efficiency is satisfactory}

b. B': {Worker's breakfast is inadequate}

c. $A \cap B'$: {Worker's efficiency is unsatisfactory and worker's breakfast is inadequate}

EXAMPLE 3.11 Refer to Example 3.10. Calculate the probabilities of the following compound events:

a. $P(A \cup B)$ b. $P(A \cap B)$ c. $P(A')$

d. $P(A' \cap B')$ e. $P(A \cup B')$

Solution

a. $P(A \cup B) = .23 + .29 + .28 = .80$

b. $P(A \cap B) = .23$

c. $P(A') = .28 + .20 = .48$

d. $P(A' \cap B') = .20$

e. $P(A \cup B') = .23 + .29 + .20 = .72$

EXAMPLE 3.12 Two hundred federal income tax returns from a particular area were classified according to the annual income and age of head of household. The results are shown below:

		Income		
		Less than $10,000	$10,000 - $30,000	Over $30,000
	Under 25	48	4	8
Age	25 - 45	28	12	56
	Over 45	4	4	36

One of the 200 returns will be randomly selected, and the following events defined:

G: {Head of household is at least 25 years old}
H: {Income is over $30,000}

Define the following events in terms of the problem and compute their probabilities.

a. G' b. $G \cup H$ c. $G \cap H'$

Solution

a. G': {Head of household is under 25 years old};

$$P(G') = \frac{48}{200} + \frac{4}{200} + \frac{8}{200} = \frac{60}{200} = .30$$

(Note that the entries within the table are counts and must be converted to probabilities. In this example, there are 60 tax returns out of 200 which satisfy event G'; thus, $P(G') = 60/200 = .30$.)

b. $G \cup H$: {Head of household is at least 25 years old, or income is over $30,000, or both};

$$P(G \cup H) = \frac{28}{200} + \frac{12}{200} + \frac{56}{200} + \frac{4}{200} + \frac{4}{200} + \frac{36}{200} + \frac{8}{200} = \frac{148}{200}$$
$$= .74$$

c. $G \cap H'$: {Head of household is at least 25 years old, and income is $30,000 or less};

$$P(G \cap H') = \frac{28}{200} + \frac{12}{200} + \frac{4}{200} + \frac{4}{200} = \frac{48}{200} = .24$$

Probability

EXERCISES

3.5 A listing of 50 houses for sale by a realty firm produced the following breakdown on number of bedrooms and bathrooms:

		Bedrooms		
		2	3	4 or More
Bathrooms	1	4	7	0
	2	5	13	11
	3 or more	0	4	6

One of these houses is to be selected at random, and the following events are defined:

A: {House has at least two bathrooms}
B: {House has two or three bedrooms}

Define the following events in terms of the problem and compute their probabilities:

a. A' b. $A \cup B$ c. $A \cap B$

d. $A \cap B'$ e. $A' \cup B$

3.6 Five hundred fourth-year medical students who responded to a nationwide survey were categorized according to their sex and their attitude toward federally-funded abortions for indigent women. The results are shown below:

		Attitude	
		Positive	Negative
Sex	Male	230	60
	Female	85	125

Suppose a single medical student is selected at random from those who responded to the survey, and the following events are defined:

A: {Student selected is male}
B: {Student has a negative attitude toward federally-funded abortions for indigent women}

Define the characteristics implied by the following compound events and compute their probabilities:

a. $A \cup B$ b. $A \cap B$ c. A'

d. $A' \cap B'$ e. $(A \cap B)'$

3.5 Conditional Probability

EXAMPLE 3.13 The 500 adult residents of a small town are categorized by age and employment status, with the following results:

		Employment Status		
		Employed	Unemployed Less than 6 Months	Unemployed At Least 6 Months
Age	Under 40	125	75	25
	40 or over	150	50	75

One of the adult residents of this town is selected at random.

a. Given that the person selected is 40 or over, what is the probability he or she is employed?

b. What is the probability that the person selected is under 40, given the subject has been unemployed at least six months?

c. Given that the person selected is unemployed, what is the probability the subject is under 40?

Solution

a. Define the following events of interest:

A: {Adult selected is employed}
B: {Adult selected is 40 or over}

Then the required probability is

$$P(A \mid B) = \frac{P(A \cap B)}{P(B)} = \frac{150/500}{(150 + 50 + 75)/500} = \frac{150}{275} \approx .55$$

b. Define

D: {Subject has been unemployed at least six months}

We desire $P(B' \mid D)$, where B is as defined above. Then

$$P(B' \mid D) = \frac{P(B' \cap D)}{P(D)} = \frac{25/500}{(25 + 75)/500} = \frac{25}{100} = .25$$

c. $$P(B' \mid A') = \frac{P(B' \cap A')}{P(A')} = \frac{(75 + 25)/500}{(75 + 25 + 50 + 75)/500} = \frac{100}{225} \approx .44$$

EXAMPLE 3.14 Refer to Example 3.4. The epidemiologists wish to compare the conditional probability that a miner has symptoms of anthracosis given that he has been in the mines less than 3 years with the conditional probability that a miner has symptoms of the disease given that he has been in the mines at least 3 years. Compute the required probabilities.

Solution We first define the following events:

A: {Miner has symptoms of anthracosis}
B: {Miner has been in the mines less than 3 years}

Then, in terms of this notation, we wish to compare $P(A \mid B)$ and $P(A \mid B')$. The required calculations are as follows:

$$P(A \mid B) = \frac{P(A \cap B)}{P(B)} = \frac{.18}{.18 + .24} = \frac{.18}{.42} \approx .43$$

$$P(A \mid B') = \frac{P(A \cap B')}{P(B')} = \frac{.52}{.52 + .06} = \frac{.52}{.58} \approx .90$$

We observe that the probability a miner with at least 3 years' experience in the mines will have symptoms of anthracosis (.90) is more than twice the probability a miner with less than 3 years' experience will have the symptoms of the disease (.43).

EXERCISES

3.7 A local business conducted a survey to determine how their customers first became acquainted with the store. The results are tabulated by sex in the following table:

		Source of Initial Information	
		Media Advertising (TV, Radio, Newspaper)	Other (Friends, Visit to Store, Etc.)
Sex	Male	36%	12%
	Female	34%	18%

A customer is to be selected randomly from the surveyed group, and the following events are defined:

A: {Customer selected is male}
B: {Customer's first acquaintance with store was through media advertising}

a. Compute $P(A \mid B)$.

b. Compute $P(B \mid A)$.

3.8 Refer to Example 3.12.

   a. What is the probability the annual income is $30,000 or less, given that the head of household is less than 25 years old?

   b. Given that the annual income is at least $10,000, what is the probability the head of household is at least 25 years old?

3.6 The Multiplicative Rule and Independent Events

EXAMPLE 3.15 The following experiment is to be performed: A fair six-sided die will be tossed once and we shall observe the number of dots showing face up. Define the following events:

A: {Observe an even number} = {2, 4, 6}
B: {Observe a 1, 2, 3, or 4} = {1, 2, 3, 4}

Are events A and B independent?

Solution Note that $P(A) = \frac{3}{6} = \frac{1}{2}$, $P(B) = \frac{4}{6} = \frac{2}{3}$,

and $P(A \mid B) = \frac{P(A \cap B)}{P(B)} = \frac{P\{2, 4\}}{P(B)} = \frac{\frac{1}{3}}{\frac{2}{3}} = \frac{1}{2}$

Since $P(A \mid B) = P(A)$, we conclude that events A and B are independent.

EXAMPLE 3.16 Past records kept on the Dow Jones Index show that on Mondays, the Index increases 55% of the time. During the remainder of the week, the Index increases on 60% of the days when it has increased the previous day, but it increases on only 30% of the days when the previous day's trading has resulted in a decrease of the Index. What is the probability that next Tuesday's trading results in an increase in the Dow Jones Index?

Solution Define the following events:

A: {The Index increases next Monday}
B: {The Index increases next Tuesday}

We wish to find $P(B)$, and are given the following information:

$P(A) = .55$, $P(B \mid A) = .6$, and $P(B \mid A') = .3$

Probability   37

Now note that B may be written as the union of two mutually exclusive events, $A \cap B$ and $A' \cap B$. (A Venn diagram may be useful to visualize this.) Thus, since the probability of the union of mutually exclusive events is equal to the sum of the probabilities of the respective events, we have

$$P(B) = P(A \cap B) + P(A' \cap B)$$
$$= P(B \mid A)P(A) + P(B \mid A')P(A')$$
(by the Multiplicative Rule of Probability)
$$= (.6)(.55) + (.3)(.45)$$
[Note that $P(A') = 1 - P(A) = 1 - .55 = .45$]
$$= .465$$

Thus, the probability is .465 that the Dow Jones Index will increase next Tuesday.

EXAMPLE 3.17 A dairy is planning to introduce two new products to the market: chocolate-flavored buttermilk and bacon-flavored ice cream. Previous consumer tests have indicated that chocolate-flavored buttermilk will be approved by the public with probability .6 and bacon-flavored ice-cream will be approved with probability .1. (Assume that the success or failure of one product is not affected by the success or failure of the other.) What is the probability that at least one of the new products will be approved by the public?

Solution We define the following events of interest:

A: {Chocolate-flavored buttermilk will be approved by the public}
B: {Bacon-flavored ice-cream will be approved by the public}

It is known that $P(A) = .6$ and $P(B) = .1$; we wish to compute $P(A \cup B)$. By the Additive Rule of Probability, we have

$$P(A \cup B) = P(A) + P(B) - P(A \cap B) = .6 + .1 - P(A \cap B)$$

Now, since events A and B are assumed to be independent, we may write

$$P(A \cap B) = P(A)P(B) = (.6)(.1) = .06$$

Finally, substitution into the Additive Rule yields

$$P(A \cup B) = .6 + .1 - .06 = .64$$

There is a 64% chance that at least one of the new products will be approved by the public.

EXERCISES

3.9 Suppose C and D are events such that $P(C) = 1/2$, $P(D) = 1/4$, and $P(C \mid D) = 1/3$.

    a. Compute $P(C \cap D)$ and $P(C \cup D)$.

    b. Are events C and D independent?

    c. Are events C and D mutually exclusive?

3.10 In the die toss experiment, define the following events:

    A: {Observe a 4}
    B: {Observe a 4, 5, or 6}
    C: {Observe an odd number}

    a. Are events A and C mutually exclusive?

    b. Are events A and B independent?

    c. Are events B and C independent?

3.11 The probability that an automobile salesman sells a new car for the sticker price is .05. If he fails to make a sale at this price, he will reduce the price of the car. He then has a 20% chance of making a sale. Find the probability the salesman makes a sale to any given customer.

3.12 Which of the following statements is (are) always true?

Let A and B be any events. Then:

    a. $P(A) + P(B) = 1$ b. $P(A) + P(A') = 1$

    c. $P(A \cup B) = P(A) + P(B)$ d. $P(A \cap B) = P(A)P(B)$

    e. $P(A \cap B) = P(B \mid A)P(A)$ f. $P(A \cap B) = 0$

    g. $P(A \mid B) = P(B \mid A)$

3.13 Which of the following statements is (are) always true?

    a. If A and B are independent events, each with nonzero probabilities, then A and B cannot be mutually exclusive.

    b. If A and B are mutually exclusive events, then A' and B' are mutually exclusive events.

    c. If $P(A \cup B) = 1$ and $P(A \cap B) = 0$, then $B = A'$.

    d. If $P(A) = P(B)$, then $P(A \mid B) = P(B \mid A)$.

3.7 Probability and Statistics: An Example

EXAMPLE 3.18 The number of typographical errors appearing in any issue of a certain independent weekly college newspaper is reported by the staff to have a mound-shaped distribution with a mean of 9 errors and a standard deviation of 2 errors. A major advertiser in the newspaper observed that last week's edition contained at least 13 typographical errors, and this week's edition contains at least 11 errors. What can the advertiser conclude?

Solution Let us define the following events:

A_1: {Last week's edition contains 13 or more errors}
A_2: {This week's edition contains 11 or more errors}

It is of interest to compute $P(A_1 \cap A_2)$, the probability of the advertiser's observing at least 13 errors in last week's edition and at least 11 errors in this week's edition of the newspaper. Note that if the staff's claim is true the relative frequency distribution of the number of typographical errors is mound-shaped with $\mu = 9$ and $\sigma = 2$. Thus, $P(A_1)$ is equivalent to the shaded area in the following diagram.

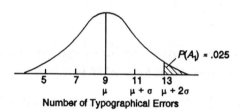

An application of the Empirical Rule for mound-shaped distributions implies

$P(A_1) \approx .025$

Similarly, we obtain

$P(A_2) \approx .16$

by relating $P(A_2)$ to the shaded area in the following diagram:

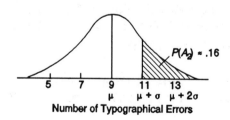

Now it is reasonable to assume that the number of typographical errors in last week's edition does not affect (and is not affected by) the number of errors in this week's edition. This assumption of independence allows us to compute $P(A_1 \cap A_2)$ by multiplying the probabilities of the respective events. Thus,

$$P(A_1 \cap A_2) = P(A_1)P(A_2) \approx (.025)(.16) = .004$$

In other words, the probability of observing at least 13 errors in last week's edition and at least 11 errors in this week's edition is only approximately .004, if the newspaper staff's claim is true. Thus, the advertiser's results would constitute a very rare (unlikely) event, if the staff's claim were true; the advertiser may thus conclude that the staff's claim regarding the relative frequency distribution of the number of typographical errors is not valid.

3.8 Random Sampling

EXAMPLE 3.19 A state's Department of Health Services (DHS) is interested in establishing a program to provide medical, financial, and emotional assistance for unwed teenaged mothers. In order to determine the types of services that would be beneficial to the subjects, DHS wishes to interview 50 of the 1,000 unwed teenagers who gave birth in this state last year.

   a. Explain how a random number table could be used to select a random sample of size 50 from the population of interest.

   b. Use Table I in the appendix of the text to identify the first 10 subjects who would be selected by the procedure described in part a.

Solution

   a. To select a random sample of 50 unwed teenagers who gave birth in the state last year, we first number the 1,000 such individuals from 000 to 999. Then we select an arbitrary starting point in a table of random numbers, such as Table I in the appendix of the text. Fifty random numbers (each of length three digits) would be obtained by proceeding from the starting point in some direction (horizontally, vertically, or diagonally). The numbers would be recorded and individuals assigned these values would be included in the sample to be interviewed.

   b. Let us begin in Row 58, Column 2, of Table I and proceed downward. Then the first 10 individuals to be selected for interviews would be those with the following number assigned:

642	763	194	333	332
009	124	080	407	578

Probability

EXERCISE

3.14 Suppose you wish to select a random sample of 100 students who have graduated from the state university system within the past three years. Explain how you could accomplish this goal.

3.9 Some Counting Rules (Optional)

EXAMPLE 3.20 Automobile license plates in a certain state are designed to show three alphabetic characters, followed by a three-digit number.

How many distinct license plates can be issued?

Solution We wish to determine the number of samples of $k = 6$ elements that can be formed by taking one element from each of 6 sets. Thus, the Multiplicative Rule is applicable, and the following $k = 6$ sets are identified:

    Set 1:   Alphabetic characters available for first position of license plate.
    Set 2:   Alphabetic characters available for second position of license plate.
    Set 3:   Alphabetic characters available for third position of license plate.
    Set 4:   Numeric digits available for fourth position of license plate.
    Set 5:   Numeric digits available for fifth position of license plate.
    Set 6:   Numeric digits available for sixth position of license plate.

The numbers of elements in the sets are $n_1 = 26$, $n_2 = 26$, $n_3 = 26$, $n_4 = 10$, $n_5 = 10$, and $n_6 = 10$. Hence, the number of different license plates which may be issued in this state is:

$$n_1 n_2 n_3 n_4 n_5 n_6 = (26)(26)(26)(10)(10)(10) = 17{,}576{,}000$$

EXAMPLE 3.21 Refer to Exercise 3.20. Suppose that the first alphabetic character must be either A, B, C, D, or E to indicate in which of the state's five administrative districts the license plate was issued. In addition, assume that the first numeric digit cannot be zero. With these restrictions, how many distinct license plates may be issued?

Solution Again, we apply the Multiplicative Rule to determine how many samples of $k = 6$ elements can be formed by taking one element from each of the 6 sets described in Example 3.20. However, since the first alphabetic character and the first numeric digit are restricted, we now have: $n_1 = 5$, $n_2 = 26$, $n_3 = 26$, $n_4 = 9$, $n_5 = 10$, $n_6 = 10$. Therefore, the number of distinct license plates is reduced to:

$$n_1 n_2 n_3 n_4 n_5 n_6 = (5)(26)(26)(9)(10)(10) = 3{,}042{,}000$$

EXAMPLE 3.22 Find the numerical values of

    a. P_2^6 b. P_0^4

Solution   a.   In general, the number of permutations of N elements taken n at a time is given by

$$P = \frac{N!}{(N-n)!}$$

In this example, we have $N = 6$ and $n = 2$; hence,

$$P_2^6 = \frac{6!}{(6-2)!} = \frac{6!}{4!} = \frac{6 \cdot 5 \cdot 4 \cdot 3 \cdot 2 \cdot 1}{4 \cdot 3 \cdot 2 \cdot 1} = 30$$

b. $P_0^4 = \dfrac{4!}{(4-0)!} = \dfrac{4!}{4!} = \dfrac{4 \cdot 3 \cdot 2 \cdot 1}{4 \cdot 3 \cdot 2 \cdot 1} = 1$

EXAMPLE 3.23   A college tennis team has 10 members. In next week's tournament against an in-state rival, five singles matches (Match #1, Match #2, etc.) will be played. In how many ways can the coach assign 5 team members to the 5 singles matches?

Solution   We are given a set of $N = 10$ team members, from which the coach is required to select $n = 5$ members for assignment to the 5 singles matches. The number of permutations of 10 elements taken 5 at a time is given by:

$$P_5^{10} = \frac{10!}{(10-5)!} = \frac{10!}{5!} = \frac{10 \cdot 9 \cdot 8 \cdot 7 \cdot 6 \cdot 5 \cdot 4 \cdot 3 \cdot 2 \cdot 1}{5 \cdot 4 \cdot 3 \cdot 2 \cdot 1} = 30{,}240$$

There are 30,240 possibilities for the assignment of players to the five matches.

EXAMPLE 3.24   A pre-school which has been experimenting with the "learning center" concept of elementary education currently has four learning centers: Counting, Alphabet, Art, and Creative Play. The first three centers are designed to accommodate four students at a time, while participation at the Creative Play center is unlimited. In how many ways can a pre-school teacher assign 18 children to the learning centers? (Assume that the Counting, Alphabet, and Art centers are each to receive four students.)

Solution   In this example, we wish to partition $N = 18$ students into $k = 4$ sets, with $n_1 = 4$, $n_2 = 4$, $n_3 = 4$, $n_4 = 18 - (4 + 4 + 4) = 6$. The number of different ways to do this is given by the Partitions Rule:

$$\frac{N!}{n_1! n_2! n_3! n_4!} = \frac{18!}{4! 4! 4! 6!}$$

$$= \frac{18 \cdot 17 \cdot 16 \cdot 15 \cdot 14 \cdot 13 \cdot 12 \cdot 11 \cdot 10 \cdot 9 \cdot 8 \cdot 7 \cdot 6 \cdot 5 \cdot 4 \cdot 3 \cdot 2 \cdot 1}{4 \cdot 3 \cdot 2 \cdot 1 \cdot 4 \cdot 3 \cdot 2 \cdot 1 \cdot 4 \cdot 3 \cdot 2 \cdot 1 \cdot 6 \cdot 5 \cdot 4 \cdot 3 \cdot 2 \cdot 1}$$

$$= 643{,}242{,}600$$

EXAMPLE 3.25 Ten members of the Statistics faculty at a certain university have expressed an interest in teaching during the summer session. However, course enrollments will require only seven teachers. In how many ways can the department chairman choose seven faculty members for summer teaching?

Solution The department chairman must select $n = 7$ faculty members from $N = 10$ available for summer employment. Since the order in which the faculty members are selected is unimportant, the number of different selections may be computed by using the Combinations Rule.

$$\binom{N}{n} = \binom{10}{7} = \frac{10!}{7!(10-7)!} = \frac{10!}{7!3!}$$

$$= \frac{10 \cdot 9 \cdot 8 \cdot 7 \cdot 6 \cdot 5 \cdot 4 \cdot 3 \cdot 2 \cdot 1}{7 \cdot 6 \cdot 5 \cdot 4 \cdot 3 \cdot 2 \cdot 1 \cdot 3 \cdot 2 \cdot 1} = 120$$

EXAMPLE 3.26 Refer to Example 3.25. Professors Jones and Smith are two of the 10 faculty members who desire to teach during the summer session. If the department chairman makes the selection of summer teachers at random, what is the probability that Professors Jones and Smith will both be selected?

Solution The experiment consists of selecting 7 faculty members from a group of 10 who desire to teach during the summer. From Example 3.25, it is known that the experiment has 120 simple events; since the selection of teachers is to be made randomly by the department chairman, each simple event is equally likely, and we have:

P(each simple event) $= 1/120$

Now we will define the event of interest as follows:

A: {Professors Jones and Smith will be selected for summer teaching}

To compute $P(A)$, it is necessary to know the number of simple events in A. Note that, in order for event A to occur, Professors Jones and Smith must both be selected, along with five of the remaining eight candidates. This may be visualized as follows:

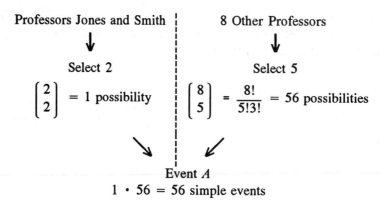

The probability of the event of interest is now easily obtained:

$$P(A) = \frac{\text{Number of simple events in } A}{\text{Total number of simple events}} = \frac{56}{120} \approx .467$$

There is an approximate 46.7% chance that both Jones and Smith will be selected for summer teaching.

EXERCISES

3.15 Find the numerical values of:

a. P_4^4 b. P_3^7 c. $\binom{9}{3}$

d. $\binom{9}{6}$ e. $\binom{10}{0}$

3.16 From a group of 14 elected representatives to a city's Citizens' Advisory Council, the mayor is going to appoint an individual to head the Budget Committee, one to head the Land Use Committee, and one to head the Parks and Recreation Committee. In how many ways may the mayor make the selections? (Assume that no individual will be appointed to head more than one committee.)

3.17 A screening procedure at a local clinic has identified 12 young adult males with elevated levels of triglyceride in their blood. In how many ways can the clinic director choose 8 of these men to participate in a diet and exercise regimen designed to reduce triglyceride levels?

3.18 Nine workers in an industrial park live in the same neighborhood and wish to establish car pools for travel to and from work. If three car pools, each containing three workers, are to be organized, in how many ways may the nine workers be assigned to the car pools?

3.19 Refer to Examples 3.25 and 3.26. What is the probability that Professor Jones, or Professor Smith, or both, will be selected for summer teaching?

3.20 Refer to Exercise 3.17. Four of the 12 young men also have a history of familial heart disease. What is the probability that at least two of these four men will be among the 8 selected for participation in the program?

Probability

CHAPTER FOUR

Discrete Random Variables

Summary

Numerical measurements of experimental phenomena are observed values of **random variables**. This chapter presented a general discussion of the characteristics of **discrete** and **continuous** random variables, and identified two discrete variables of particular interest: the **binomial** and **Poisson** random variables.

Knowledge of the **probability distribution** of a random variable allows one to calculate the probabilities of specific sample observations. In addition, approximate probability statements about the behavior of a random variable may be based on numerical descriptive measures (mean and standard deviation) of the probability distribution.

4.1 Two Types of Random Variables

EXAMPLE 4.1 Classify the following random variables as discrete or continuous. Specify the possible values the random variables may assume.

a. x = the number of customers who enter a particular bank during the noon hour on a given day.

b. x = the time (in seconds) required for a teller to serve a bank customer.

c. x = the distance (in miles) between a randomly-selected home in a community and the nearest pharmacy.

d. x = the number of traffic accidents involving school buses during a week in a particular state.

e. x = the number of tosses of a fair coin required to observe at least three heads in succession.

Solution

a. discrete, $x = 0, 1, 2, \ldots$

b. continuous; $0 < x < \infty$

c. continuous; $0 < x < \infty$

d. discrete; $x = 0, 1, 2, \ldots$

e. discrete; $x = 3, 4, 5, 6, \ldots$

EXAMPLE 4.2 Which of the following describe discrete random variables and which describe continuous random variables?

    a. The number of raisins in a 16 oz. package of raisin bran.

    b. The number of shares of stock traded on the New York Stock Exchange the day after a presidential State of the Union message.

    c. The volume (in cubic feet) of heated space in a randomly selected suburban shopping mall.

    d. The number of applications for FM radio station licenses received in a week by the Federal Communications Commission.

    e. The distance a randomly selected steel-belted radial tire will be driven before it develops a defect.

    f. The time between emergency shut-downs at a nuclear power plant.

Solution   discrete: a, b, d   continuous: c, e, f

EXERCISES

4.1 Classify each of the following random variables as discrete or continuous. Specify the possible values the random variables may assume.

    a. x = the number of small businesses in the metropolitan Miami area which will make a profit during the next fiscal year.

    b. x = the number of burglary reports received at a local police department during a holiday weekend period.

    c. x = the diameter (in centimeters) of a randomly selected ball bearing produced by a machining process.

    d. x = the total amount of rainfall received during the week of the World Series in the two participating cities.

    e. x = the length (in miles) of the daily delivery route of a randomly selected Postal Service employee.

    f. x = the number of complaints received by the customer service representatives of a large department store during a given week.

4.2 Give an example of a discrete random variable that may be of interest in each of the following areas:

   a. Sociology   b. Psychology

   c. Political Science d. Biology

   e. Chemistry   f. Business

4.2 Probability Distributions for Discrete Random Variables

EXAMPLE 4.3 Determine whether each of the following represents a valid probability distribution. If not, explain why not.

a.
x	$p(x)$
0	.20
1	.90
2	−.10

b.
x	$p(x)$
−2	.3
−1	.3
1	.3
2	.3

c.
x	$p(x)$
−1	.25
0	.65
1	.10

Solution

a. Not valid; $p(2)$ cannot be negative.

b. Not valid; $\sum_{\text{All } x} p(x)$ must equal 1.

c. Valid.

EXAMPLE 4.4 It is known that 80% of all tomato plants sold at a local garden store bear tomatoes within 45 days. You have just purchased two tomato plants from the store and are interested in the random variable x = the number of the two plants which will produce tomatoes within 45 days. Construct the probability distribution for x, assuming the growth patterns of the two purchased plants are independent.

Solution It is first required to list the simple events associated with this experiment: (T_1, T_2), (T_1, F_2), (F_1, T_2) and (F_1, F_2), where (T_1, F_2) indicates the first plant produces **tomatoes** within 45 days and the second plant **fails** to produce tomatoes within 45 days, etc.

The random variable x assigns to each simple event the numerical value equal to the number of plants which will produce tomatoes within 45 days. (Note that x may assume the values 0, 1, and 2.) Thus, we have:

Simple Event	Value of x Assigned to Simple Event
(T_1, T_2)	2
(T_1, F_2)	1
(F_1, T_2)	1
(F_1, F_2)	0

Now to compute probabilities associated with each value of x, we note

$$P(x = 0) = p(0) = P(F_1, F_2) = P(F_1)P(F_2) \quad \text{(by independence)}$$
$$= (.2)(.2) = .04$$

$$P(x = 1) = p(1) = P(T_1, F_2) + P(F_1, T_2) = P(T_1)P(F_2) + P(F_1)P(T_2)$$
$$= (.8)(.2) + (.2)(.8) = .32$$

$$P(x = 2) = p(2) = P(T_1, T_2) = P(T_1)P(T_2)$$
$$= (.8)(.8) = .64$$

Finally, the probability distribution for x is written as follows:

x	$p(x)$
0	.04
1	.32
2	.64

EXAMPLE 4.5 The following is the probability distribution for x = the number of bedrooms in a randomly selected home listed for sale in a particular region of the country:

x	$p(x)$
1	.05
2	.15
3	.30
4	.42
5	.08

Find the probability that a randomly selected home listed for sale in this region will have:

a. At least 3 bedrooms.

b. Fewer than 3 bedrooms.

Discrete Random Variables

c. No more than 4 bedrooms.

d. Exactly 1 bedroom.

Solution

a. $P(x \geq 3) = p(3) + p(4) + p(5) = .30 + .42 + .08 = .80$

b. $P(x < 3) = p(1) + p(2) = .05 + .15 = .20$

c. $P(x \leq 4) = p(1) + p(2) + p(3) + p(4) = .05 + .15 + .30 + .42 = .92$

d. $P(x = 1) = p(1) = .05$

EXERCISES

4.3 The following table gives the probability distribution for a random variable x:

x	$p(x)$
0	.422
1	.422
2	.141
3	?

a. Compute the value of $p(3)$ for this random variable.

b. Find $P(x \leq 2)$.

c. Find $P(x > 1)$.

4.4 Past records of an educational institution indicate that 15% of the students who receive federally-insured loans default within the first three years after graduation. We will randomly select two loan recipients who graduated in 1981 and record x = the number of people (in our sample of two) who default on their loans.

a. Construct the probability distribution for the random variable x.

b. What is the probability that at least one of the selected students will default within three years after graduation?

4.3 Expected Values of Discrete Random Variables

EXAMPLE 4.6 Refer to the following probability distribution for a random variable x:

x	$p(x)$
−1	.1
0	.1
1	.2
2	.2
5	.4

a. Compute the mean μ and the standard deviation σ for this random variable.

b. Specify the interval $\mu \pm 2\sigma$; what is the probability that x will fall within the interval $\mu \pm 2\sigma$?

Solution a. $E(x) = \mu = \sum_{\text{All } x} xp(x)$
$= -1(.1) + 0(.1) + 1(.2) + 2(.2) + 5(.4)$
$= -.1 + 0 + .2 + .4 + 2$
$= 2.5$

Note that $\mu = 2.5$ does not represent a possible value of x. Rather, in many repetitions of the experiment, the **average** value of x which will be observed is 2.5.

Now we will calculate σ^2, the variance of x:

$\sigma^2 = E[(x - \mu)^2] = \sum_{\text{All } x} (x - \mu)^2 p(x)$
$= (-1 - 2.5)^2(.1) + (0 - 2.5)^2(.1) + (1 - 2.5)^2(.2) + (2 - 2.5)^2(.2)$
$\quad + (5 - 2.5)^2(.4)$
$= (-3.5)^2(.1) + (-2.5)^2(.1) + (-1.5)^2(.2) + (-.5)^2(.2) + (2.5)^2(.4)$
$= (12.25)(.1) + (6.25)(.1) + (2.25)(.2) + (.25)(.2) + (6.25)(.4)$
$= 1.225 + .625 + .450 + .050 + 2.500$
$= 4.85$

Thus, the standard deviation is $\sigma = \sqrt{4.85} \approx 2.20$.

b. $\mu \pm 2\sigma = 2.5 \pm 2(2.20) = 2.5 \pm 4.40$, or $(-1.9, 6.9)$

We observe that the two standard deviation interval about the mean will contain **all** observations on the random variable x, i.e.,

$$P(-1.9 \le x \le 6.9) = 1$$

EXAMPLE 4.7 A risky investment involves paying $100 that will return either $900 (for a net profit of $800) with probability .2 or $0 (for a net loss of $100) with probability .8. What is your expected net profit from this investment?

Solution Let x = net profit from this investment. We wish to compute $E(x)$, where x has the following probability distribution:

x	$p(x)$
800	.2
−100	.8

(Note that a loss is treated as a negative profit.) Then

$$E(x) = \sum_{\text{All } x} xp(x) = 800(.2) + (-100)(.8) = 160 - 80 = 80$$

Your expected net profit on an investment of this type is $80. If you were to make a very large number of such investments, some would result in a net profit of $800, and others would result in a net loss of $100. However, in the long run, your **average** net profit per investment would be $80.

EXAMPLE 4.8 From past experience, an automobile insurance company knows that a given automobile will suffer a total loss with probability .02, a 50% loss with probability .08, or a 25% loss with probability .15 during a year. What annual premium should the company charge to insure a $10,000 automobile, if it wishes to "break even" on all policies of this type? (Assume there will be no other partial loss.)

Solution The company will break even if it charges a premium equal to the **average** payoff on each policy. Thus, we let x = payoff on a policy of this type, and note that x has the following probability distribution:

x	$p(x)$	
$10,000	.02	(represents a total loss)
$5,000	.08	(represents a 50% loss)
$2,500	.15	(represents a 25% loss)
$0	.75	

(Note that with probability .75, the automobile will incur no loss, and hence the company will make no payoff on the policy.) Now,

$$E(x) = \$10{,}000(.02) + \$5{,}000(.08) + \$2{,}500(.15) + \$0(.75) = \$975$$

That is, the company will make an **average** payoff of $975 on each policy of this type. In order to break even, a premium of $975 should be assessed.

EXERCISES

4.5 The probability distribution of x = the number of siblings for high school students attending a large preparatory school is as follows:

x	$p(x)$
0	.11
1	.19
2	.37
3	.23
4	.08
5	.02

a. Compute the mean, variance, and standard deviation of the random variable x.

b. What is the probability that a randomly observed value of x will fall within the interval $\mu \pm 2\sigma$?

4.6 A popular automobile advertisement claims that a certain car seats 2 adults and 2.3 children comfortably, and thus is perfect for the "average" American family. Comment on the statistical sensibility of this advertisement.

4.4 The Binomial Random Variable

EXAMPLE 4.9 For each of the following experiments, decide whether x is a binomial random variable.

a. From past records, it is known that 5% of all electronic calculators manufactured by a certain company will need major repairs within 3 months. Your company has just purchased 10 calculators from this firm. Let x be the number of these calculators which will require major repairs within 3 months.

b. Let x be the number of tosses of a fair coin before the first head is observed.

c. Five of the members of a governor's Council on Youth Fitness are female, and five are male. Three of the ten members of this committee will be selected randomly to appear before the state legislature to request funds to support physical education in the elementary schools. Let x be the number of females chosen.

d. Past experience indicates that 1% of mass-marketed paperback books contain at least one production error (e.g., missing or misplaced sections). A book store has just purchased 1,500 copies of a recently released paperback. Let x be the number of these paperbacks which contain at least one production error.

Solution

a. If we are willing to assume that the 10 calculators were randomly selected from all those produced by the company, and that they operate independently, then x is a binomial random variable with $n = 10$ and $p = .05$, where $p = $ the probability a randomly selected calculator requires major repairs within 3 months.

b. The binomial model is not satisfactory since n, the number of identical trials to be performed, cannot be determined in advance.

c. The trials do not satisfy the assumption of independence. To see this, suppose the first member selected is female. Then the probability that the second member selected is female decreases from 1/2 to 4/9, since only four of the nine remaining members are female. Hence, x is not a binomial random variable.

d. The random variable x is binomial with $n = 1,500$ and $p = .01$, where p is the probability that a randomly selected paperback will contain at least one production error.

EXAMPLE 4.10 Past studies have shown that 20% of all surgical patients who receive a particular type of anesthetic experience mild post-operative nausea. Suppose that four patients are scheduled to receive this type of anesthetic today, and let $x = $ the number who will experience post-operative nausea. Tabulate the probability distribution for x.

Solution

We first note that x satisfies the characteristics of a binomial random variable with $n = 4$ and $p = .2$. Then the probability distribution for this random variable is given by

$$p(x) = \binom{n}{x} p^x q^{n-x} = \binom{4}{x}(.2)^x(.8)^{n-x}$$

Probabilities are then computed as follows:

$$p(0) = \binom{4}{0}(.2)^0(.8)^4 = \frac{4!}{0!4!}(.2)^0(.8)^4$$

$$= \frac{4 \cdot 3 \cdot 2 \cdot 1}{1 \cdot 4 \cdot 3 \cdot 2 \cdot 1}(1)(.4096) = .4096$$

[Note that $0! = 1$ and $(.2)^0 = 1$.]

$$p(1) = \frac{4!}{1!3!}(.2)^1(.8)^3 = \frac{4 \cdot 3 \cdot 2 \cdot 1}{1 \cdot 3 \cdot 2 \cdot 1}(.2)^1(.8)^3 = 4(.2)(.512) = .4096$$

$$p(2) = \frac{4!}{2!2!}(.2)^2(.8)^2 = 6(.04)(.64) = .1536$$

$$p(3) = \frac{4!}{3!1!}(.2)^3(.8)^1 = 4(.008)(.8) = .0256$$

$$p(4) = \frac{4!}{4!0!}(.2)^4(.8)^0 = 1(.0016)(1) = .0016$$

Thus, the probability distribution of x is:

x	$p(x)$
0	.4096
1	.4096
2	.1536
3	.0256
4	.0016

EXAMPLE 4.11 Manufactured items that do not pass inspection are often sold as "seconds" or "blemishes" at a reduced price. Quite often, such a product may have only a minor defect which does not affect performance. Past testing has shown that, for a particular product, 90% of all "seconds" perform as well at "firsts." A random sample of 25 "seconds" of this product is to be selected. We will record x = the number of items in the sample which perform as well as "firsts."

a. Find $P(x \geq 20)$.

b. Find $P(18 \leq x < 23)$.

c. Compute μ and σ for the random variable x.

d. If this experiment were to be repeated many times, what proportion of the x observations would fall within the interval $\mu \pm 2\sigma$?

Solution a. We first observe that x is a binomial random variable with $n = 25$ and $p = .9$, where p is the probability a randomly selected "second" will perform as well as a "first," or, equivalently, the proportion of "seconds" which perform as well as "firsts." We now refer to Table II in the appendix to the text to find the desired probabilities.

$$P(x \geq 20) = 1 - P(x \leq 19) = 1 - .033 = .967.$$

b. $P(18 \leq x < 23) = P(x \leq 22) - P(x \leq 17) = .463 - .002 = .461$

c. For a binomial random variable x with parameters n and p, $\mu = np$ and $\sigma^2 = npq$. Thus, for our example, $\mu = 25(.9) = 22.5$, $\sigma^2 = 25(.9)(.1) = 2.25$, and $\sigma = \sqrt{2.25} = 1.5$.

d. The two standard deviation interval about the mean is

$$\mu \pm 2\sigma = 22.5 \pm 2(1.5) = 22.5 \pm 3.0, \text{ or } (19.5, 25.5)$$

Now, $P(19.5 \leq x \leq 25.5) = P(x \geq 20)$, since the largest possible value x may assume is 25. From the table of binomial probabilities,

$$P(x \geq 20) = 1 - P(x \leq 19) = 1 - .033 = .967$$

Essentially all the observations on this random variable will fall within two standard deviations of the mean.

EXAMPLE 4.12 Assume x is a binomial random variable with $n = 15$ and $p = .6$. Use the table of binomial probabilities to find:

a. $P(x \leq 12)$

b. $P(x < 11)$

c. $P(x > 6)$

d. $P(x \geq 5)$

e. $P(4 < x < 12)$

f. $P(4 \leq x \leq 12)$

g. $P(6 \leq x < 12)$

Solution

a. $P(x \leq 12) = .973$

b. $P(x < 11) = P(x \leq 10) = .783$

c. $P(x > 6) = 1 - P(x \leq 6) = 1 - .095 = .905$

d. $P(x \geq 5) = 1 - P(x \leq 4) = 1 - .009 = .991$

e. $P(4 < x < 12) = P(x \leq 11) - P(x \leq 4) = .909 - .009 = .900$

f. $P(4 \leq x \leq 12) = P(x \leq 12) - P(x \leq 3) = .973 - .002 = .971$

g. $P(6 \leq x < 12) = P(x \leq 11) - P(x \leq 5) = .909 - .034 = .875$

EXERCISES

4.7 For each of the following experiments, decide whether x is a binomial random variable:

a. A car dealer has 20 used cars, half of which are domestic models and half of which are foreign models. For a weekend special sale, he chooses 6 of these cars at random and reduces their price. Let x be the number of domestic models selected for the sale.

b. During the past year, 25% of the tobacco crops in a certain state were contaminated by an insecticide which had been improperly manufactured. The state's Secretary of Agriculture will examine 60 randomly selected tobacco crops in the state and will record x, the number which have been contaminated by the insecticide.

c. One judge in a traffic court is known to revoke the licenses of 80% of the drivers charged with failure to yield the right of way. Fourteen drivers who have been charged with this violation are scheduled to appear before this judge in tomorrow's court session. Let x be the number of these drivers whose licenses will be revoked.

4.8 A restaurant owner has observed that 30% of the dinner bills are paid with a major credit card. Three bills are randomly chosen during a given evening, and the method of payment will be noted. Let x be the number of these bills which are paid with a major credit card.

   a. Tabulate the probability distribution for the random variable x.

   b. What is the probability that at least two of the bills will be paid with a major credit card?

4.9 Records kept by a newspaper publisher in a major city indicate that 40% of all subscriptions are for the morning paper. A sample of 15 subscriptions is to be randomly selected from the publisher's records. Let x be the number of these subscriptions which are for the morning paper.

   a. Find $P(x < 8)$.

   b. Find $P(6 \leq x \leq 11)$.

   c. Find $P(x \geq 5)$.

   d. Compute the mean and variance of the random variable x.

   e. If this experiment were to be repeated many times, what proportion of the x observations would fall within the interval $\mu \pm 2\sigma$?

4.5 The Poisson Random Variable (Optional)

EXAMPLE 4.13 A switchboard operator has observed that an average of three calls for emergency assistance arrive during each 4-hour shift. Let x be the number of such calls which arrive during a given shift. Specify the probability distribution of the random variable x.

Solution The random variable x has the characteristics of a Poisson distribution, since it represents the number of independent events (calls for emergency assistance) occurring over a fixed time period (4-hour shift), with a fixed average ($\lambda = 3$). Thus, the probability distribution for x is:

$$p(x) = \frac{\lambda^x e^{-\lambda}}{x!} = \frac{3^x e^{-3}}{x!} \text{ for } x = 0, 1, 2, \ldots$$

EXAMPLE 4.14 The number, x, of daily breakdowns of an obsolete university computer has an average of 1.5.

   a. What is the probability that there will be no breakdowns on a particular day?

   b. What is the probability there will be at least two breakdowns on a given day?

Discrete Random Variables

c. What is the probability the computer has no breakdowns for two consecutive days?

d. Specify μ and σ^2 for the random variable x.

Solution We first note that x possesses the characteristics of a Poisson random variable with $\lambda = 1.5$; thus, the probability distribution is given by:

$$p(x) = \frac{\lambda^x e^{-\lambda}}{x!} = \frac{(1.5)^x e^{-1.5}}{x!}, \text{ for } x = 0, 1, 2, \ldots$$

a. $P(x = 0) = \dfrac{(1.5)^0 e^{-1.5}}{0!} = \dfrac{1 \cdot e^{-1.5}}{1} = .223$

b. $P(x \geq 2) = P(x = 2) + P(x = 3) + P(x = 4) + \ldots$
$= 1 - [P(x = 0) + P(x = 1)]$
$= 1 - \left[\dfrac{(1.5)^0 e^{-1.5}}{0!} + \dfrac{(1.5)^1 e^{-1.5}}{1!} \right]$
$= 1 - [.223 + (1.5)(.223)]$
$= 1 - (.558) = .442$

c. Define the following events:

A: {No breakdown on Day 1}
B: {No breakdown on Day 2}

Now, if we assume independence of the computer's operation from day to day, we have

$$P(A \cap B) = P(A)P(B) = (.223)(.223) = .050$$

There is only a 5% chance that the computer will operate for two consecutive days without a breakdown.

d. For a Poisson random variable, $\mu = \lambda$ and $\sigma^2 = \lambda$. In our example, $\mu = \sigma^2 = 1.5$.

EXERCISES

4.10 An insurance company claims they receive an average of five calls per hour reporting automobile accidents.

a. What is the probability that either 4 or 5 accidents are reported during the next hour?

b. What is the probability that at least one accident is reported during the next hour?

4.11 Experiments with a new recording technique indicate that an average of two defects occur on each side of a phonograph record.

   a. What is the probability that no defects occur on a particular side of a given record?

   b. What is the probability that at least two defects occur on each side of a particular record?

CHAPTER FIVE

Continuous Random Variables

Summary

This chapter discussed the methodology used to describe **continuous** random variables, and presented the probability distributions of three such variables with wide applicability in statistics: the **normal, uniform** and **exponential** distributions. It was also demonstrated that the normal probability distribution provides a good approximation for the binomial distribution when the sample size is sufficiently large.

5.1 Continuous Probability Distributions
5.2 The Uniform Distribution

EXAMPLE 5.1 A machine which is designed to produce bolts with a diameter of .50 inch has been found to produce bolts with diameters which are distributed uniformly between .48 and .52 inch.

    a. What fraction of the bolts produced by the machine are at least .49 inch in diameter?

    b. Compute the mean and standard deviation of bolt diameters.

    c. Find the probability that the diameter of a randomly selected bolt lies within two standard deviations of the mean.

Solution   a. Let x be the diameter of a randomly selected bolt produced by the machine. The probability distribution of x is known to be uniform over the interval $.48 \leq x \leq .52$; thus, $c = .48$, $d = .52$, and

$$f(x) = \frac{1}{.52 - .48} = \frac{1}{.04} = 25 \quad (.48 \leq x \leq .52)$$

This uniform probability function may be visualized as a rectangle with length .04 and height 25, as shown at the right.

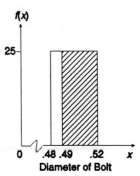

Now, since the area of a rectangle is equal to the base times the height, we have

$$P(x \geq .49) = (.52 - .49) \times 25 = .75$$

Three-fourths of the bolts produced by the machine are at least .49 inch in diameter.

b. For a uniform random variable,

$$\mu = \frac{c + d}{2}, \text{ and } \sigma = \frac{d - c}{\sqrt{12}}$$

In our example,

$$\mu = \frac{.48 + .52}{2} = .50, \text{ and } \sigma = \frac{.52 - .48}{\sqrt{12}} = .012$$

The diameters of the bolts produced by this machine have a mean of .50 inch and a standard deviation of .012 inch.

c. The interval $\mu \pm 2\sigma$, or $.50 \pm 2(.012)$, or $(.476, .524)$ covers the entire range of the distribution. Thus, the diameters of all bolts produced will lie within two standard deviations of the mean.

EXERCISES

5.1 A receptionist for a dentist has observed from long experience that patients show up anytime from 15 minutes early to 30 minutes late for their appointments. Assume that the distribution of patient arrival times is uniform over the interval between −15 and 30, with 0 representing the scheduled appointment time.

   a. Find the probability that the dentist's next patient is not late for his or her appointment.

   b. What fraction of all patients arrive within five minutes of their scheduled appointment time?

5.2 Consumer investigations have shown that at auto service centers, jobs for which the customer is given an estimate of one hour labor will actually require anywhere between 30 and 70 minutes. Assume the distribution of actual labor times is uniform over this interval.

   a. If a customer is charged in advance for an estimated labor of one hour, what is the probability this will represent an excessive charge in terms of the actual labor time required for the job?

   b. Compute the mean and standard deviation of the actual labor time required on jobs for which the customer is given an estimate of one hour.

Continuous Random Variables

5.3 The Normal Distribution

EXAMPLE 5.2 Use Table IV in the appendix of the text to find the following probabilities relating to the standard normal random variable, z.

a. $P(0 \leq z \leq 1.86)$
b. $P(-.52 \leq z \leq 0)$

c. $P(-.30 \leq z \leq 1.76)$
d. $P(z \leq 1.25)$

e. $P(z \geq 2.08)$
f. $P(1.23 \leq z \leq 1.94)$

Solution It is often helpful to draw a sketch of the normal curve to assist in using Table IV to find the required probabilities.

a. Note that Table IV is designed to provide areas (probabilities) of the form $P(0 \leq z \leq z_0)$ for specified values of z_0. Thus, to determine $P(0 \leq z \leq 1.86)$, find the tabled entry at the intersection of the row headed 1.8 and the column headed .06. You will find $P(0 \leq z \leq 1.86) = .4686$.

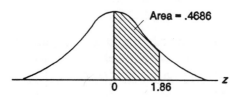

b. Because of the symmetry of the normal distribution, the area between $-.52$ and 0 is equal to the area between 0 and .52.

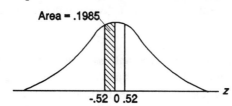

Thus, $P(-.52 \leq z \leq 0) = P(0 \leq z \leq .52) = .1985$

c. $P(-.30 \leq z \leq 1.76) = P(-.30 \leq z \leq 0) + P(0 \leq z \leq 1.76)$
$= .1179 + .4608 = .5787$

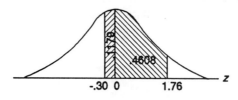

d. Note from Table IV that the area between 0 and 1.25 is .3944. However, to include **all** of the area to the left of 1.25, we write

$$P(z \leq 1.25) = P(z \leq 0) + P(0 \leq z \leq 1.25)$$
$$= .5 + .3944 = .8944$$

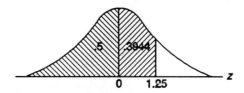

e. To find the required upper-tail area, we note that, since the area between 0 and 2.08 is .4812, the area to the right of 2.08 is $.5 - .4812 = .0188$. Thus,

$$P(z \geq 2.08) = .0188$$

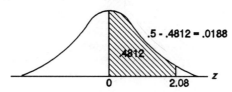

f. To find the area between 1.23 and 1.94, we observe that the area between 0 and 1.94 is $P(0 \leq z \leq 1.94) = .4738$. However, we want **only** the area between 1.23 and 1.94, so it is required to subtract from .4738 the area between 0 and 1.23, giving

$$P(1.23 \leq z \leq 1.94) = .4738 - .3907 = .0831$$

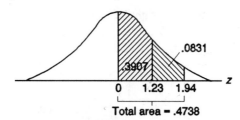

EXAMPLE 5.3 Find the value of z_0 which makes the following statements true:

a. $P(0 \leq z \leq z_0) = .3749$ b. $P(-z_0 \leq z \leq z_0) = .9500$

c. $P(z \geq z_0) = .0495$

Continuous Random Variables

Solution

a. We know that the area between 0 and z_0 is .3749. Since Table IV is set up to provide areas of this form, it is required to find the entry .3749 in the body of the table. You will find .3749 at the intersection of the 1.1 row and the .05 column; thus,

$$P(0 \leq z \leq 1.15) = .3749 \text{ and } z_0 = 1.15$$

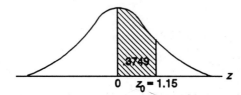

b. We use the symmetry of the normal distribution to conclude that the area within a distance of z_0 on each side of 0 is .9500/2 = .4750.

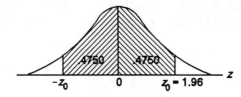

Thus, $P(0 \leq z \leq z_0) = .4750$ and $z_0 = 1.96$.

c. The first step is to determine on which side of 0 the value z_0 should be. Since $P(z \geq 0) = .5$, and we want a smaller probability, z_0 should be to the right of 0. [Note that for $z_0 < 0$, the value of $P(z \geq z_0)$ would be greater than .5, which is too large.]

Since the area to the right of z_0 is $P(z \geq z_0) = .0495$, this leaves an area of .5 − .0495 = .4505 between 0 and z_0. The location of .4505 in the body of Table IV implies $z_0 = 1.65$.

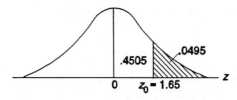

EXAMPLE 5.4 The time required to perform a certain laboratory test to analyze blood chemistry is approximately normally distributed with a mean of 90 seconds and a standard deviation of 10 seconds.

a. What is the probability that a randomly selected blood sample takes at least 105 seconds to analyze?

b. What fraction of the blood samples take between 80 and 95 seconds to analyze?

Solution   a. Let x = the time required for analysis of a blood sample. Since x has an approximately normal distribution, we can transform to the standard normal distribution to compute the required probabilities. The z-score corresponding to an analysis time of $x = 105$ seconds is

$$z = \frac{x - \mu}{\sigma} = \frac{105 - 90}{10} = 1.5$$

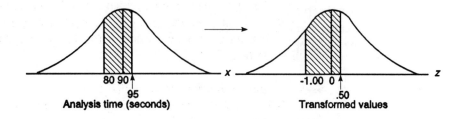

Thus, $P(x \geq 105) = P(z \geq 1.5) = .5 - .4332 = .0668$. Almost 7% of all blood samples require at least 105 seconds for analysis.

b. $P(80 < x < 95) = P\left(\dfrac{80 - 90}{10} < z < \dfrac{95 - 90}{10}\right)$
$= P(-1.00 < z < .50)$
$= .3413 + .1915$
$= .5328$

Approximately 53% of all blood samples require between 80 and 95 seconds for analysis.

EXAMPLE 5.5   The guarantees associated with consumer products must be carefully determined. The manufacturer wants to set the guarantee so that the product looks very attractive, but so that very few items will have to be replaced because of failure before the expiration of the guarantee.

Tests on new steel-belted radial tires showed an average tire wear of 40,000 miles and a standard deviation of 3,000 miles. If tire wear is assumed to be approximately normally distributed, how much tire wear should be guaranteed if the manufacturer wishes to replace only 1% of the tires sold?

Continuous Random Variables

$\mu = 40,000$
$\sigma = 3,000$

Solution If we let x = the tire wear for a randomly selected tire and G = the amount of tire wear guaranteed, we can write

$$P(x < G) = .01$$

since if the amount of tire wear is less than what the manufacturer guarantees, the tire must be replaced. The next step is to determine how many standard deviations G is from the mean by transforming the normal random variable x to the standard normal random variable z:

$$z = \frac{G - \mu}{\sigma} = \frac{G - 40,000}{3,000}$$

Now, since we know the area to the left of z (it is equivalent to the area to the left of G), we can use Table IV to determine the value of z (and hence G).

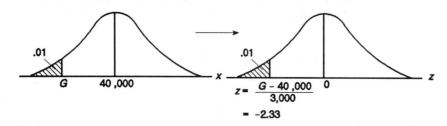

We observe that the area to the left of G (.01) corresponds to the area beneath the z-curve to the left of -2.33, since $P(z < -2.33) \approx .01$. Therefore, we can solve the equation

$$\frac{G - 40,000}{3,000} = -2.33$$

which results in

$$G = 40,000 - 2.33(3,000) = 33,010$$

By setting the guarantee at approximately 33,000 miles, the manufacturer will need to replace approximately 1% of all tires sold.

EXERCISES

5.3 Use Table IV in the appendix of the text to find the following probabilities:

a. $P(z \geq 1.75)$

b. $P(-2.02 \leq z \leq -1.44)$

c. $P(z \leq -1.38)$

d. $P(-1.48 \leq z \leq 1.03)$

5.4 Find the value of z_0 which makes the following statements true:

a. $P(-z_0 \leq z \leq z_0) = .9902$

b. $P(z \leq z_0) = .0250$

c. $P(z \geq z_0) = .7734$

5.5 Television ratings are based on the percentage of all homes with televisions which are tuned in to a show. For nightly prime-time viewing, ratings are normally distributed with a mean rating of 21% and a standard deviation of 4%.

a. What proportion of all nightly prime-time shows rate between 20 and 25?

b. What percentage of prime-time shows rate at least 30?

5.6 A recent report claimed that college graduates with degrees in industrial engineering have an average starting salary of $2,654 per month. Assume that the distribution of starting salaries for industrial engineers is normal, with a standard deviation of $200 per month.

a. What proportion of industrial engineers receive a starting salary of at least $2,500 per month?

b. A recent graduate claims that his starting salary is in the top 5% of starting salaries for all industrial engineers. If his claim is true, what is the minimum possible value for his monthly salary?

5.4 Approximating a Binomial Distribution with a Normal Distribution

EXAMPLE 5.6 From extensive records, a major airline has concluded that 10% of all people with confirmed reservations do not show up for their flight. For a small plane with 96 seats, the airline has decided to book 100 reservations. What is the probability that everyone who shows up with a reservation for the flight gets a seat?

Solution Let x = the number of people with a reservation who show up for the flight. Then x is a binomial random variable with $n = 100$ and $p = .9$, where p is the probability that a person holding a reservation shows up for the flight. We want to find $P(x \leq 96)$, since everyone will get a seat if 96 or fewer people show up for the flight. Binomial tables for $n = 100$ are not available in the text; thus, we will determine if the normal approximation to the binomial may be applied in this situation.

The mean, variance, and standard deviation of the binomial random variable x are computed as follows:

$$\mu = np = 100(.9) = 90$$
$$\sigma^2 = npq = 100(.9)(.1) = 9$$
$$\text{and} \quad \sigma = 3$$

Since the interval $\mu \pm 3\sigma$, or (81, 99) lies completely within the interval from 0 to $n = 100$, it is appropriate to use a normal approximation to the binomial probabilities.

To apply the continuity correction, we note that it is necessary to include all the area to the left of 96.5 beneath the approximating normal curve in order to include all of the area corresponding to $x = 96$. Thus,

$$P(x \leq 96) = P(x \leq 96.5) \approx P\left[z \leq \frac{96.5 - 90}{3}\right]$$
$$= P(z \leq 2.17) = .5 + .4850 = .9850$$

Thus, there is a 98.5% chance that all those who show up with reservations for the flight will be accommodated.

EXAMPLE 5.7 Suppose that x is a binomial random variable for which you wish to find approximate probabilities by using the normal approximation. Indicate the continuity correction which would be appropriate for each of the following situations:

a. $P(x > a)$ b. $P(x < a)$

c. $P(x \geq a)$ d. $P(x \leq a)$

e. $P(a \leq x \leq b)$ f. $P(a < x < b)$

Solution It may be useful to sketch the binomial histograms and the approximating normal curves in order to visualize each situation.

a. $P(x > a) = P(x > a + 1/2)$

b. $P(x < a) = P(x < a - 1/2)$

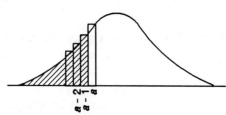

c. $P(x \geq a) = P(x \geq a - 1/2)$

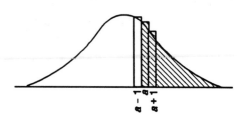

d. $P(x \leq a) = P(x \leq a + 1/2)$

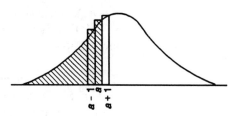

e. $P(a \leq x \leq b)$
$= P(a - 1/2 \leq x \leq b + 1/2)$

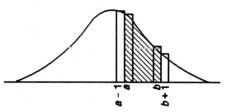

f. $P(a < x < b)$
$= P(a + 1/2 < x < b - 1/2)$

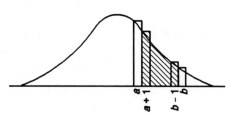

EXERCISES

5.7 A television dealer has observed that 75% of all televisions that he sells are portable. Find the approximate probability that, of the next 50 sets sold, at least 35 will be portable.

5.8 The Department of Education in a certain state wishes to establish a state-subsidized school lunch program. It claims that 40% of all elementary school children in the state are now inadequately nourished. In a sample of 150 elementary school children from this state, what is the probability that at least 50 are inadequately nourished, if the Department of Education's claim is valid?

5.5 The Exponential Distribution (Optional)

EXAMPLE 5.8 An important decision facing bank managers is how many tellers to have available for duty at different times of the day. This decision is based on several factors, one of which is the time required by tellers to service customers. A recent study at a local bank indicated that the service times of the tellers have an exponential distribution with an average service time of 4 minutes.

a. Find the probability that a customer would require at least 6 minutes to be served at this bank.

b. What proportion of this bank's customers can be served in less than 4 minutes?

c. What proportion of the service times fall within two standard deviations of the mean?

Solution a. If we let x = service time required for a randomly selected customer, then x has an exponential distribution with $\mu = 4$. The probability distribution of x is then given by

$$f(x) = \frac{1}{\theta}e^{-x/\theta} = \frac{1}{4}e^{-x/4} \quad (x > 0)$$

where $\theta = \mu = 4$.

We can now write $P(x \geq a) = e^{-a/\theta} = e^{-a/4}$. Thus,

$$P(x \geq 6) = e^{-6/4} = e^{-1.50} \approx .22$$

There is an approximate 22% chance that a customer will require at least 6 minutes to be serviced at this bank.

b. $P(x < 4) = 1 - P(x \geq 4) = 1 - e^{-4/4} = 1 - e^{-1.0} \approx 1 - .37 = .63$

Approximately 63% of the bank's customers can be served in less than 4 minutes.

c. We wish to compute the probability that an x observation falls within the interval $\mu \pm 2\sigma$, or $4 \pm 2(4)$, or $(-4, 12)$. Since x cannot assume negative values, this is equivalent to

$$P(0 < x < 12) = 1 - P(x \geq 12) = 1 - e^{-12/4} = 1 - e^{-3.0}$$
$$\approx 1 - .05 = .95$$

Thus, approximately 95% of the service times will fall within two standard deviations of the mean.

EXERCISES

5.9 The attendant at a car wash has observed that a car arrives every 5 minutes, on the average. Assume that the length of time between arrivals has an exponential distribution. A car has just arrived at the car wash.

a. What is the probability that the next car arrives in less than 2 minutes?

b. What is the probability that the next car arrives in between 1 and 5 minutes?

5.10 An advertisement claims that a smoke detector system will last for an average of two years before the batteries have to be replaced. If we assume the lifelength of the system has an exponential distribution, what proportion of all systems sold will last at least two years?

CHAPTER SIX

Sampling Distributions

Summary

The objective of most statistical investigations is to make an inference about a population **parameter**, θ. To do this, we use sample data to compute a sample **statistic** which contains information about θ. The **sampling distribution** of a statistic characterizes the distribution of values of the statistic over a very large number of samples.

Unbiasedness and **minimum variance** are desirable properties of the probability distribution of a sample statistic. In terms of these criteria, the sample mean provides the "best" estimator when the parameter of interest is the population mean, μ. Further, the **Central Limit Theorem** guarantees that the sampling distribution for the sample mean will be approximately normal, regardless of the distribution of the sampled population, when the sample size is sufficiently large.

For all the statistics used in this text, the variance of the sampling distribution of the statistic is inversely related to the sample size.

6.1 What Is a Sampling Distribution?
6.2 Properties of Sampling Distributions: Unbiasedness and Minimum Variance

EXAMPLE 6.1 Consider the random variable x, whose probability distribution is as follows:

x	1	2	3	4	5
$p(x)$	.2	.2	.2	.2	.2

Construct the sampling distribution (or probability distribution) of $\bar{x}$, the mean of a random sample of $n = 3$ observations selected without replacement from this population.

Solution We first note the form of the probability distribution for the random variable x:

In addition, you should verify that $\mu = E(x) = 3.00$.

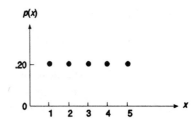

A random sample of size $n = 3$ is to be selected without replacement from this population. The 10 possible samples and their associated values of $\bar{x}$ are as follows:

Possible Sample	Value of $\bar{x}$
1, 2, 3	2.00
1, 2, 4	2.33
1, 2, 5	2.67
1, 3, 4	2.67
1, 3, 5	3.00
1, 4, 5	3.33
2, 3, 4	3.00
2, 3, 5	3.33
2, 4, 5	3.67
3, 4, 5	4.00

Since the selection is made at random, each sample is equally likely and has probability .1.

To construct the probability distribution of $\bar{x}$, we observe that

$P(\bar{x} = 2.00) = P(\text{sample } 1, 2, 3) = .1$
$P(\bar{x} = 2.33) = P(\text{sample } 1, 2, 4) = .1$
$P(\bar{x} = 2.67) = P(\text{sample } 1, 2, 5) + P(\text{sample } 1, 3, 4) = .1 + .1 = .2$
etc.

Thus, the probability distribution of $\bar{x}$ for this particular situation is given by:

$\bar{x}$	$p(\bar{x})$
2.00	.1
2.33	.1
2.67	.2
3.00	.2
3.33	.2
3.67	.1
4.00	.1

The mean of the probability distribution of $\bar{x}$ is equal to

$$\mu_{\bar{x}} = E(\bar{x}) = 2.00(.1) + 2.33(.1) + 2.67(.2) + 3.00(.2) + 3.33(.2) \\ + 3.67(.1) + 4.00(.1) \\ = 3.00$$

We have demonstrated that the mean of the sampling distribution of $\bar{x}$ is equal to the mean of the sampled population; i.e., $E(\bar{x}) = \mu$, and thus the sample mean $\bar{x}$ is an unbiased estimator of the population mean μ.

It should also be noted that the values of $\bar{x}$ cluster more closely about their mean than do the values of x; for example,

$$P(2.5 \leq x \leq 3.5) = .2$$

whereas

$$P(2.5 \leq \bar{x} \leq 3.5) = .6$$

In other words, the variation of the $\bar{x}$ values is less than the variation of the x values. It can be shown that, among all possible unbiased estimators of a population mean, $\bar{x}$ is the unbiased estimator with minimum variance.

EXERCISES

6.1 Refer to Example 6.1.

  a. Construct the sampling distribution of m, the median of a random sample of size $n = 3$ selected without replacement from the population in Example 6.1.

  b. Show that $E(m) = \mu$ (i.e., m is an unbiased estimator of μ) **in this particular situation.**

  c. Use the probability distribution for $\bar{x}$ shown in Example 6.1 to compute $\sigma_{\bar{x}}^2$. Use the probability distribution for m found in part a of this exercise to compute σ_m^2. Show that $\sigma_{\bar{x}}^2 < \sigma_m^2$; i.e., for these two unbiased estimators of the population mean, $\bar{x}$ has a smaller variance than m.

6.2 Consider the random variable x, whose probability distribution is as follows:

x	1	3	4	5	10
$p(x)$	.2	.2	.2	.2	.2

  a. Construct the sampling distribution of m, the median of a random sample of size $n = 3$ selected without replacement from this population.

  b. Verify that $E(m) \neq \mu$, and thus, the sample median is not, in general, an unbiased estimator of the population mean.

6.3 The Central Limit Theorem
6.4 The Relation Between Sample Size and a Sampling Distribution

EXAMPLE 6.2 Due to problems with foam during the filling process, beer bottles are not always filled to capacity. A certain brewery advertises that their bottles contain, on the average, 12 ounces of beer. A random sample of 100 bottles off their production line yielded a sample mean fill of 11.9 ounces, and a standard deviation of .4 ounce. Compute the probability of observing a sample mean fill of 11.9 ounces or less, assuming the brewery's claim is valid.

Sampling Distributions

Solution   Let x be the amount of fill (in ounces) of a randomly selected beer bottle from this brewery. Then x has a probability distribution (the exact form of which is unspecified) with mean $\mu = 12$, according to the brewery's claim.

We wish to compute $P(\bar{x} \leq 11.9)$, where $\bar{x}$ is the mean of a random sample of $n = 100$ observations from the distribution of x. The Central Limit Theorem assures us that $\bar{x}$ has an approximately normal distribution, with

and $\quad$ mean $\mu_{\bar{x}} = \mu = 12$

$$\text{standard deviation } \sigma_{\bar{x}} = \frac{\sigma}{\sqrt{n}} \approx \frac{s}{\sqrt{n}} = \frac{.4}{\sqrt{100}} = .04$$

(Note that the value of σ, the standard deviation of the sampled population, is unknown; hence, we estimate it using the value of s, the sample standard deviation.)

Now we apply a result from Chapter 5 to conclude that $z = \dfrac{\bar{x} - \mu_{\bar{x}}}{\sigma_{\bar{x}}}$ is a standard normal random variable. Thus,

$$P(\bar{x} \leq 11.9) = P\left(z \leq \frac{11.9 - 12}{.04}\right) = P(z \leq -2.5) = .0062$$

The probability of observing a sample mean fill of 11.9 ounces or less is only .0062, if the brewery's claim that $\mu = 12$ is true. We have strong evidence that the brewery's claim is untrue, because the observed sample result is very unlikely if the claim is true.

EXAMPLE 6.3   A study of the residential housing in a particular state showed that the average appraised value of a house in the state is $85,000 and the standard deviation is $8,000. A random sample of 40 homes is to be selected from the state.

a. Describe the sampling distribution of the sample mean appraised value of the 40 homes.

b. Compute $P(\bar{x} > \$86{,}000)$.

c. Compute $P(\$82{,}000 \leq \bar{x} \leq \$86{,}000)$.

Solution   a. Let x be the appraised value of a randomly selected home from this state. Then, the probability distribution of x (although the exact form is unknown) has mean $\mu = 85{,}000$ and standard deviation $\sigma = 8{,}000$. Now, by an application of the Central Limit Theorem, the sampling distribution of $\bar{x}$, the mean of a random sample of size $n = 40$, is approximately normal with

and $\quad$ mean $\mu_{\bar{x}} = \mu = 85{,}000$

$$\text{standard deviation } \sigma_{\bar{x}} = \frac{\sigma}{\sqrt{n}} = \frac{8{,}000}{\sqrt{40}} = 1{,}264.9$$

b. $P(\bar{x} > 86{,}000) = P\left(z > \dfrac{86{,}000 - 85{,}000}{1{,}264.9}\right) = P(z > .79) = .2148$

c. $P(82{,}000 \leq \bar{x} \leq 86{,}000) = P\left(\dfrac{82{,}000 - 85{,}000}{1{,}264.9} \leq z \leq \dfrac{86{,}000 - 85{,}000}{1{,}264.9}\right)$

$= P(-2.37 \leq z \leq .79) = .7763$

EXAMPLE 6.4 Refer to Example 6.3. Suppose it is desired to reduce the standard deviation of the sampling distribution of $\bar{x}$ to one-third of its original value, i.e., to

$$\sigma_{\bar{x}} = \frac{1{,}264.9}{3} = 421.63$$

What size sample must be selected to accomplish this?

Solution Since we desire $\sigma_{\bar{x}} = 421.63$, it is necessary to solve the following equation for n:

$$\sigma_{\bar{x}} = \frac{\sigma}{\sqrt{n}} = \frac{8{,}000}{\sqrt{n}} = 421.63$$

or

$$n = \left[\frac{8{,}000}{421.63}\right]^2 = 360$$

Note that in order to reduce $\sigma_{\bar{x}}$ to one-third of its original value, it is necessary to include nine times as many observations in the sample. In general, to reduce the standard deviation of the sampling distribution of $\bar{x}$ to $1/k$ of its original value, nk^2 observations must be included in the sample.

EXERCISES

6.3 A grocery store advertises that its ground beef contains no more than 30% fat. A random sample of 64 one-pound packages of ground beef was cooked and then weighed electronically to determine fat content. The weights of the cooked beef had an average of $\bar{x} = .68$ pound and a standard deviation of $s = .075$ pound. [Note that, if the store's advertisement is accurate, then the population mean (μ) cooked weight of all such one-pound packages sold at the store would be at least .70 pound.] Compute the probability of observing a sample mean cooked weight of .68 pound or less for these 64 packages, assuming μ does in fact equal .70 pound.

6.4 For families living in a particular housing district, the average annual income is $41,000 and the standard deviation is $15,000.

a. Describe the sampling distribution of $\bar{x}$, the sample mean annual income of $n = 25$ families selected at random from the housing district.

b. Suppose it is desired to reduce the standard deviation of the sampling distribution in part **a** by one-half. What size sample must be selected to accomplish this?

c. Compute $P(\$46{,}000 \leq \bar{x} \leq \$44{,}900)$ for each of the sampling distributions in parts **a** and **b**.

CHAPTER SEVEN

Inferences Based on a Single Sample: Estimation

Summary

This chapter presented the inference-making technique of **estimation** based on a single sample selected from a population; measures of the uncertainty of the inference, based on the sampling distribution of the sample statistic, were discussed.

An interval estimate, called a **confidence interval**, is used to estimate a population parameter with a prespecified probability of coverage, called the **confidence coefficient**.

When using the sample mean $\bar{x}$ to make inferences about the population mean μ, the standard normal z statistic is employed when the sample size is sufficiently large ($n > 30$). For small samples drawn from a normal population with an unknown value of σ, the t statistic is used in making the inference. The z statistic is also used to form confidence intervals for a binomial proportion p, based on a sample fraction of successes, $\hat{p}$.

The sample size required for estimating a population parameter can be determined by specifying the desired confidence level and the bound on the error of estimation.

7.1 Large-Sample Estimation of a Population Mean

EXAMPLE 7.1 For each of the following confidence coefficients, determine the value of $z_{\alpha/2}$ which would be used in constructing a large-sample $100(1 - \alpha)\%$ confidence interval for μ:

    a. .80   b. .85   c. .95

Solution   a.   For a confidence coefficient of .80, we have

$$1 - \alpha = .80, \text{ or } \alpha = .20, \text{ or } \alpha/2 = .10.$$

Thus, we desire the value of $z_{\alpha/2} = z_{.10}$ which locates an area of .10 in the upper tail of the distribution of z (see figure):

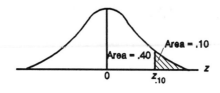

Now recall that the table in your text gives areas beneath the standard normal curve between 0 and the specified z value. From the figure, it can be seen that the area between 0 and $z_{.10}$ is $.50 - .10 = .40$; thus, it is necessary to locate the area .4000 in the body of the table in order to determine the corresponding value of $z_{.10}$. The tabled entry .3997 is the value closest to the desired area of .4000, and its corresponding z value is $z_{.10} = 1.28$. Hence the value of $z_{\alpha/2}$ used in the construction of an 80% confidence interval for μ is $z_{.10} = 1.28$.

b. For a confidence coefficient of .85,

$$1 - \alpha = .85, \text{ or } \alpha = .15, \text{ or } \alpha/2 = .075$$

The required value of z is the one which locates an upper-tail area of .075 beneath the z distribution:

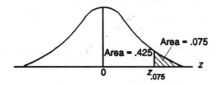

The area between 0 and the desired value of $z_{.075}$ is $.5 - .075 = .425$. The closest entry in the body of the table is .4251, and it corresponds to a z value of 1.44. In the construction of a large-sample 85% confidence interval for μ, the appropriate z value is

$$z_{.075} = 1.44$$

c. For a confidence coefficient of .95,

$$1 - \alpha = .95, \text{ or } \alpha = .05, \text{ or } \alpha/2 = .025$$

The area between 0 and $z_{\alpha/2} = z_{.025}$ is equal to $.5 - .025 = .475$.

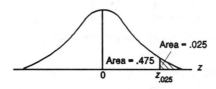

From the table of areas beneath the standard normal distribution, it is seen that the desired z value is

$$z_{.025} = 1.96$$

EXAMPLE 7.2 As part of a study designed to investigate the relationship between diet and coronary heart disease, 50 young adult males at a particular clinic were selected for initial screening and follow-up. At entry into the study, the cholesterol level was obtained for each of the 50 individuals, and the following results were obtained (values are in milligrams of cholesterol per deciliter of blood):

$$\bar{x} = 224.8$$
$$s = 18.4$$

a. Construct a 95% confidence interval for μ, the true (but unknown) mean cholesterol value for all young adult males who might have been chosen for the study.

b. Interpret the confidence interval obtained in part a.

Solution a. The general form of a large-sample 95% confidence interval for a population mean μ is

$$\bar{x} \pm z_{.025}\sigma_{\bar{x}}, \text{ or } \bar{x} \pm 1.96\left(\frac{\sigma}{\sqrt{n}}\right)$$

In our example we have

$$\bar{x} = 224.8, s = 18.4, n = 50$$

Since the value of σ (the population standard deviation of the cholesterol levels for all young adult males) is unknown, we will estimate it with the sample standard deviation, s. Then the 95% confidence interval is given by

$$224.8 \pm 1.96\frac{\sigma}{\sqrt{50}} \approx 224.8 \pm 1.96\left(\frac{18.4}{\sqrt{50}}\right) = 224.8 \pm 5.1$$

or

(219.7, 229.9)

b. If we were to select repeated random samples of 50 young adult males, compute $\bar{x}$ (the sample mean cholesterol level) for each sample, and construct a 95% confidence interval each time, then approximately 95% of the intervals so constructed would contain the true value of the population mean μ. We are thus 95% confident that the interval (219.7, 229.9) contains μ, although we do not know whether this **particular** interval is one of the 95% which contain μ, or one of the remaining 5% which fail to contain μ.

EXERCISES

7.1 For each of the following confidence coefficients, specify the value of $z_{\alpha/2}$ which would be used in the construction of a large-sample $100(1 - \alpha)\%$ confidence interval for a population mean μ.

    a. .90   b. .98   c. .99

7.2 An examination of the yearly premiums for a random sample of 80 automobile insurance policies from a major company showed an average of $429 and a standard deviation of $49.

    a. Give a precise definition of what the population parameter μ represents in terms of this problem.

    b. Construct a 99% confidence interval for μ.

    c. Interpret the interval constructed in part b.

7.3 A publisher was interested in estimating the mean retail cost of its best-selling hardbound novel. A random sample of 50 retail outlets gave the following results on retail cost:

$$\bar{x} = \$19.80$$
$$s = \$2.40$$

Estimate the mean retail cost of the publisher's best-selling hardbound novel. Use a 95% confidence interval.

7.2 Determining the Sample Size Necessary to Estimate a Population Mean

EXAMPLE 7.3 Each year, when automotive makers introduce the new models, the Environmental Protection Agency (EPA) calculates an estimate of the average gasoline mileage rating for each new model. For a particular new model, the EPA intends to provide an estimate of the average mileage rating which is accurate to within .5 mile per gallon (mpg) with 95% confidence. How many cars should be tested by EPA in order to achieve the desired accuracy? Assume that the standard deviation of mpg ratings for this model was $\sigma = 2.5$ mpg last year.

Solution The EPA desires to estimate μ, the true mean gasoline mileage rating for this model correct to within $B = .5$ mpg with 95% confidence. The required sample size is

$$n = \frac{(z_{\alpha/2})^2 \sigma^2}{B^2}$$

where $z_{\alpha/2} = z_{.025} = 1.96$, $\sigma = 2.5$, and $B = .5$. The solution is then computed as follows:

$$n = \frac{(1.96)^2 (2.5)^2}{(.5)^2} = 96.04$$

Thus, the EPA must test at least 97 automobiles to estimate μ with the desired accuracy.

EXERCISE

7.4 A problem facing the United States Postal Service is the rather large number of letters and parcels which cannot be delivered because of an illegible or nonexistent address. The Postal Service would like to estimate μ, the daily mean number of items which cannot be delivered in a particular region. For how many randomly selected days should the Postal Service record the number of undeliverable items in order to estimate μ to within 6 items with 95% confidence? (Previous records indicate that the standard deviation of the daily number of undeliverable items in this area is 25.)

7.3 Small-Sample Estimation of a Population Mean

EXAMPLE 7.4 Use the table of critical values of the t distribution to find the particular values of t_0 which make the following statements true:

a. $P(t > t_0) = .05$ when df = 12

b. $P(t < t_0) = .025$ when df = 20

c. $P(t > t_0 \text{ or } t < -t_0) = .01$ when df = 8

Solution
a. We first note that the table of critical values of the t distribution in your text gives values t_α such that $P(t > t_\alpha) = \alpha$. Now, at the intersection of the column labeled $t_{.05}$ and the row corresponding to 12 degrees of freedom, we find the entry 1.782; thus,

$$P(t > 1.782) = .05 \text{ when df} = 12$$

(see figure), i.e., $t_0 = 1.782$.

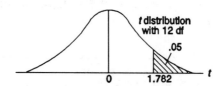

b. From the table, we observe that $P(t > 2.086) = .025$ when df = 20:

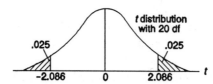

Now we use the symmetry of the t distribution to conclude that

$$P(t < -2.086) = .025 \text{ when df} = 20,$$

i.e., $t_0 = -2.086$.

c. We wish to locate the critical values t_0 and $-t_0$ such that the total area in the two tails of the t distribution with 8 degrees of freedom is .01:

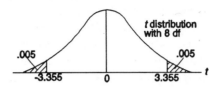

Because of the symmetry of the t distribution, an area of $.01/2 = .005$ in each tail is required. Thus, we determine the value of t_0 by locating the entry at the intersection of the column labeled $t_{.005}$ and the row corresponding to df = 8; $t_0 = 3.355$. Thus,

$$P(t > 3.355 \text{ or } t < -3.355) = .01 \text{ when df} = 8$$

EXAMPLE 7.5 With inflation and rising operational costs, medical bills have risen rapidly over the past 10 years. A newspaper study recently reported that the nationwide average cost of a hospital room in a semi-private ward is $350 per day. To test this claim, an insurance company conducted its own survey in California, the only state in which it is licensed to sell hospitalization policies. The daily cost of a semi-private room was recorded for a random sample of 20 major hospitals, and the following statistics were computed: $\bar{x} = \$336$; $s = \$23$. Construct a 98% confidence interval for the true mean cost of a hospital room in a semi-private ward.

Solution The general form of a small-sample confidence interval for μ is

$$\bar{x} \pm t_{\alpha/2}(s/\sqrt{n})$$

where t is based on $n - 1$ degrees of freedom. In our example, $\bar{x} = 336$, $s = 23$, $n = 20$, and $t_{\alpha/2} = t_{.01} = 2.539$, based on 19 df. Thus, the required confidence interval is

$$336 \pm 2.539(23/\sqrt{20})$$
or $\quad 336 \pm 2.539(5.14)$
or $\quad 336 \pm 13.05$
or $\quad (322.95, 349.05)$

The insurance company can be 98% confident that the true mean cost of a hospital room in a semi-private ward is between $322.95 and $349.05.

EXAMPLE 7.6 A trans-oceanic airline conducted a study to estimate the mean weight of baggage checked by passengers on its Miami to London flight. A random sample of 25 passengers was selected, and the weight of each passenger's checked baggage was recorded. The following results were obtained: $\bar{x} = 43.5$ pounds; $s = 6$ pounds. Construct a 95% confidence interval for the true mean weight of checked baggage of passengers on the Miami to London flight.

Solution   The general form of a small-sample confidence interval for μ is

$$\bar{x} \pm t_{\alpha/2}(s/\sqrt{n}),$$

where t is based on $n - 1$ degrees of freedom. In our example, $\bar{x} = 43.5$, $s = 6$, $n = 25$, and $t_{\alpha/2} = t_{.025} = 2.064$, based on 24 df. Thus, the required confidence interval is

$$43.5 \pm 2.064(6/\sqrt{25})$$
or $\quad 43.5 \pm 2.064(1.2)$
or $\quad 43.5 \pm 2.48$
or $\quad (41.02, 45.98)$

The airline can be 95% confident that the true mean weight of checked baggage of passengers going from Miami to London is between 41.02 and 45.98 pounds.

Note that this confidence interval procedure requires the assumption that the relative frequency distribution of baggage weights for passengers on the Miami to London flight is approximately normal.

EXERCISES

7.5 Determine the values of t_0 that make the following statements true:

a. $P(t < t_0) = .01$ when df $= 18$

b. $P(t > t_0) = .05$ when df $= 7$

c. $P(t < -t_0 \text{ or } t > t_0) = .10$ when df $= 22$

7.6 The off-campus housing office at a major university has published an apartment-finder's guide which states that the average cost of a two-bedroom, unfurnished apartment located within three miles of the campus is $500 per month. However, for a random sample of 16 apartment complexes contacted by a prospective renter, the mean and standard deviation of the monthly rental rates for such apartments were found to be $520 and $28, respectively.

a. Construct a 99% confidence interval for the mean cost of a two-bedroom, unfurnished apartment within three miles of the campus.

b. State any assumptions required for the validity of the test procedure used in part a.

7.7 A random sample of 22 new homes built last year in a particular state was selected, and the number of square feet of heated floor space was recorded for each. The sample mean and standard deviation were computed as follows: $\bar{x} = 1,712$; $s = 250$.

a. Construct a 95% confidence interval for the mean square footage of new homes being built in this state.

b. State any assumptions required for the validity of the procedures used in part a.

7.4 Large-Sample Estimation of a Binomial Probability

EXAMPLE 7.7 Each year, the Internal Revenue Service (IRS) conducts a study of the accuracy of tax returns for the purpose of possible simplification (or other revision) of forms for subsequent tax years. In this year's study, a random sample of 500 tax returns showed that 90 contained at least one error. Construct a 99% confidence interval for p, the proportion of all tax returns that contain at least one error.

Solution The general form of a large-sample confidence interval for p is

$$\hat{p} \pm z_{\alpha/2}\sqrt{pq/n} \approx \hat{p} \pm z_{\alpha/2}\sqrt{\hat{p}\hat{q}/n},$$

where $\hat{p}$ is the proportion of successes in the sample, and $\hat{q} = 1 - \hat{p}$.

In this example, we are interested in $p =$ the true proportion of all tax returns submitted this year that contain at least one error. The best estimate of the binomial parameter p is

$$\hat{p} = \frac{\text{Number of returns in sample that contain at least one error}}{\text{Number of returns in sample}}$$

$$= \frac{90}{500} = .18$$

Now, we substitute the values

$$\hat{p} = .18, z_{\alpha/2} = z_{.005} = 2.58, \hat{q} = 1 - \hat{p} = .82, n = 500$$

into the general formula to obtain the desired confidence interval:

$$.18 \pm 2.58\sqrt{(.18)(.82)/500} = .18 \pm 2.58(.017)$$
$$= .18 \pm .044 \text{ or } (.136, .224)$$

The IRS can be 99% confident that the proportion of this year's tax returns that contain at least one error lies within the interval (.136, .224).

EXAMPLE 7.8 In previous years, a mail-order company observed that 10% of its customers placed an additional order within six months of their original order. However, the records for a random sample of 1,000 recent customers indicate that only 80 customers placed an additional order within six months of their original order. Construct a 95% confidence interval for p, the proportion of customers that place an additional order within six months.

Solution The general form of a large-sample confidence interval for p is

$$\hat{p} \pm z_{\alpha/2}\sqrt{pq/n} \approx \hat{p} \pm z_{\alpha/2}\sqrt{\hat{p}\hat{q}/n}$$

where $\hat{p}$ is the proportion of successes in the sample, and $\hat{q} = 1 - \hat{p}$.

We are interested in $p =$ the true proportion of customers who place an additional order within six months. The best estimate of p is

$$\hat{p} = \frac{80}{1,000} = .08$$

Now we substitute the values $\hat{p} = .08$, $\hat{q} = 1 - \hat{p} = .92$, $z_{\alpha/2} = z_{.025} = 1.96$, $n = 1,000$ into the general formula to obtain the desired confidence interval:

$$.08 \pm 1.96\sqrt{(.08)(.92)/1,000} = .08 \pm 1.96(.009)$$
$$= .08 \pm .018 \text{ or } (.062, .098)$$

The mail-order company can be 95% confident that the true proportion of customers who place additional orders within six months is between .062 and .098.

EXERCISES

7.8 A local dairy is considering the possibility of ceasing home deliveries in an effort to minimize cost increases for its products. In a survey of 100 randomly selected customers who currently receive home delivery, 64 customers indicated that they were in favor of this economy measure.

Construct a 99% confidence interval for the proportion of all home delivery customers who favor the proposed measure.

7.9 Refer to Exercise 7.8. The dairy will cease home deliveries if the results of the survey indicate (at $\alpha = .01$) that at least 60% of all home delivery customers favor the proposal. What decision should the dairy make?

7.5 Determining the Sample Size Necessary to Estimate a Binomial Probability

EXAMPLE 7.9 A state senator is interested in estimating p, the proportion of his constituents in favor of reinstating the military draft. How many voters must be sampled in order to produce an estimate of p which is correct to within .02 with 90% confidence?

Solution The required sample size is

$$n = \frac{(z_{\alpha/2})^2 pq}{B^2}$$

where $z_{\alpha/2} = z_{.05} = 1.645$ and $B = .02$. Since we have no prior knowledge about the value of p, we will substitute $p = q = .5$ into the sample size formula. (This is a conservative procedure which will yield a value for n which is at least as large as required.)

Substitution yields:

$$n = \frac{(1.645)^2(.5)(.5)}{(.02)^2} = 1{,}691.3$$

The senator must sample 1,692 voters in order to estimate p with the desired accuracy.

EXAMPLE 7.10 A national retail chain wishes to estimate p, the proportion of charge customers who are more than one month behind in their payments. If they want to be 95% confident that their estimate is within .01 of the true value of p, how many accounts should be sampled? (Past evidence indicates that the proportion of delinquent accounts is approximately .15.)

Solution The required sample size is

$$n = \frac{(z_{\alpha/2})^2 pq}{B^2}$$

where $z_{\alpha/2} = z_{.025} = 1.96$, $B = .01$, and we use our prior estimate of $p = .15$ ($q = .85$) in the computation. Thus,

$$n = \frac{(1.96)^2(.15)(.85)}{(.01)^2} = 4{,}898.04$$

The retail chain must sample the accounts of 4,899 charge customers in order to be 95% confident that their estimate is within .01 of the true value of p.

EXERCISE

7.10 A manufacturer of desk calculators believes that the proportion of calculators which require service within one month of sale is no more than .04. The firm's quality control engineer wishes to estimate p, the actual proportion of all calculators sold by the firm which require service within one month of sale, accurate to within .005 with 90% confidence. The sale/repair records for how many calculators should be sampled in order to obtain the desired accuracy?

Inferences Based on a Single Sample: Estimation

CHAPTER EIGHT

Inferences Based on a Single Sample: Tests of Hypotheses

Summary

This chapter presented the inference-making technique of hypothesis testing based on a single sample selected from the population; measures of the uncertainty of the inference, based on the sampling distribution of the sample statistic, were discussed.

Two methods for reporting the results of a hypothesis test were presented. In the first method, the experimenter fixes the probability α of falsely rejecting the **null hypothesis** in favor of the **research (alternative) hypothesis**. The second method reports the **observed significance level**, or *p*-value, which is the extent to which the test statistic disagrees with the null hypothesis, and leaves to the reader the decision whether to reject the null hypothesis.

When using the sample mean $\bar{x}$ to perform a test of hypothesis about the population mean μ, the standard normal z statistic is employed when the sample size is sufficiently large ($n > 30$). For small samples drawn from a normal population with an unknown σ, the t statistic is used to perform the hypothesis test. The z statistic is also used to perform a test of hypothesis for a binomial proportion p, based on the sample fraction of successes, $\hat{p}$.

8.1 The Elements of a Test of Hypothesis
8.2 Large-Sample Test of a Hypothesis About a Population Mean

EXAMPLE 8.1 A college recruiter claimed that the average monthly starting salary for 1992 college graduates with business degrees would be $2,500. The monthly starting salaries for a sample of 75 companies selected randomly from the placement files were recorded, with the following results: $\bar{x} = \$2,481$; $s = \$91$. Is this evidence that the average starting salary is less than $2,500 per month? Use a significance level of $\alpha = .05$.

Solution The elements of this large-sample test of a hypothesis are as follows:

The null hypothesis is

H_0: $\mu = 2,500$

where μ is the true (but unknown) mean monthly starting salary (in dollars) for all 1992 college graduates with degrees in business.

The alternative (or research) hypothesis that we wish to establish is

$$H_a: \mu < 2{,}500$$

i.e., the average starting salary is less than $2,500 per month.

The rejection region consists of all values of z such that

$$z < -z_\alpha$$
or $\quad z < -z_{.05}$
or $\quad z < 1.645$

Thus, we will reject H_0 if the computed value of the test statistic is less than -1.645.

The test statistic is

$$z = \frac{\bar{x} - \mu_0}{\sigma_{\bar{x}}} = \frac{\bar{x} - \mu_0}{\sigma/\sqrt{n}}$$

where μ_0 is the value assigned to μ in the null hypothesis, and the value of σ will be estimated by s. Thus,

$$z = \frac{2{,}481 - 2{,}500}{91/\sqrt{75}} = -1.81$$

Since this value of the test statistic falls within the rejection region, we reject H_0. There is sufficient evidence to conclude that the mean starting salary is less than $2,500 per month. We recognize the possibility of having made a Type I error; if the null hypothesis is in fact true, the probability that we have incorrectly rejected H_0 is $\alpha = .05$.

EXAMPLE 8.2 A local pizza parlor advertises that its average time for delivery of a pizza is within 30 minutes of receipt of the order. The delivery times for a random sample of 64 orders were recorded, with the following results: $\bar{x} = 34$ minutes; $s = 21$ minutes. Is there sufficient evidence to conclude that the actual mean delivery time is larger than what is claimed by the pizza parlor? Use a significance level of $\alpha = .01$.

Solution The hypothesis test is composed of the following elements:

$$H_0: \mu = 30$$
$$H_a: \mu > 30$$

where μ is the true mean delivery time (in minutes) for all orders placed at the pizza parlor. Note that we are interested in establishing the alternative hypothesis that the true mean delivery time exceeds the value advertised by the pizza parlor.

The rejection region consists of all values of z such that

$$z > z_\alpha$$
or $\quad z > z_{.01}$
or $\quad z > 2.33$

The null hypothesis H_0 will be rejected if the computed value of the test statistic z exceeds 2.33.

The test statistic is

$$z = \frac{\bar{x} - \mu_0}{\sigma_{\bar{x}}} = \frac{\bar{x} - \mu_0}{\sigma/\sqrt{n}} = \frac{34 - 30}{21/\sqrt{64}} = 1.52$$

This value of the test statistic does not fall within the rejection region; thus, there is insufficient evidence to conclude that the true mean delivery time is significantly greater than the 30 minutes claimed by the pizza parlor.

EXAMPLE 8.3 A machine is designed to fill cereal boxes with a net weight of 16 ounces. It is important that the machine operate accurately: if it fills too much, the company wastes excess cereal; if it underfills the boxes, the company risks a penalty from the Food and Drug Administration. The company has instituted a new quality control program to monitor the amount of fill of its cereal boxes. Every four hours, a random sample of 100 boxes is selected from the production line, and the amounts of fill are noted. If there is evidence (at $\alpha = .05$) that the mean amount of fill differs from 16 ounces, then the filling machine is reset.

Suppose one such inspection yielded the following results: $\bar{x} = 15.98$ ounces; $s = .21$ ounce. Should the machine be reset?

Solution We wish to perform a test of the hypothesis

$$H_0: \mu = 16$$

against

$$H_a: \mu \neq 16$$

where μ is the true mean amount of fill (in ounces) of the cereal boxes. We perform a two-tailed test to be able to detect the possibility that the machine either underfills (i.e., $\mu < 16$) or overfills (i.e., $\mu > 16$) the cereal boxes.

The rejection region consists of the following sets of z values:

$$z < -z_{\alpha/2} \quad \text{or} \quad z > z_{\alpha/2}$$
i.e., $\quad z < -z_{.025} \quad \text{or} \quad z > z_{.025}$
i.e., $\quad z < -1.96 \quad \text{or} \quad z > 1.96$

The test statistic is

$$z = \frac{\bar{x} - \mu_0}{\sigma/\sqrt{n}} \approx \frac{\bar{x} - \mu_0}{s/\sqrt{n}} = \frac{15.98 - 16}{.21/\sqrt{100}} = -.95$$

Since the value of the test statistic does not fall within the rejection region, we do not reject H_0. There is insufficient evidence to conclude that the mean amount of fill differs significantly from 16 ounces; there is no need to reset the machine based on this inspection.

EXERCISES

8.1 A random sample of 49 plastic bags with an advertised breaking strength of 10 pounds were tested, yielding an average breaking strength of 9.7 pounds and a standard deviation of .7 pound. Does this experiment provide sufficient evidence (at $\alpha = .01$) to conclude that the true mean breaking strength of all plastic bags is less than the manufacturer's claim of 10 pounds?

8.2 A manufacturing process was designed to make pistons with a mean diameter of 11.5 centimeters. In the most recent quality control test of the process, a random sample of 81 pistons was selected from the production line, and their diameters were measured, with the following results: $\bar{x} = 11.54$ centimeters; $s = .25$ centimeter. Do the test results suggest (at $\alpha = .05$) that the mean diameter of the pistons being produced differs from 11.5 centimeters?

8.3 Observed Significance Levels: *p*-Values

EXAMPLE 8.4 For the given hypotheses and computed value of the test statistic, determine the observed significance level of the large-sample test.

a. H_0: $\mu = 50$, H_a: $\mu > 50$; $z = 1.79$

b. H_0: $\mu = 75$, H_a: $\mu < 75$; $z = -2.04$

c. H_0: $\mu = 4.8$, H_a: $\mu \neq 4.8$; $z = 1.32$

Solution a. Since the test is upper-tailed (H_a: $\mu > 50$), values of the test statistic even more contradictory to H_0 than the computed value would be values larger than 1.79. Thus, the observed significance level (*p*-value) for this test is

$$p = P(z \geq 1.79) = .5 - .4633 = .0367$$

b. The test is lower-tailed (H_a: $\mu < 75$); values of the test statistic that are contradictory to H_0 are those in the lower tail of the distribution. The observed significance level (*p*-value) for this test is

$$p = P(z \leq -2.04) = .5 - .4793 = .0207$$

Inferences Based on a Single Sample: Tests of Hypotheses

c. For this two-tailed test (H_a: $\mu \neq 4.8$), values of the test statistic at least as contradictory to the null hypothesis as the computed value are those less than -1.32 **or** greater than 1.32. Thus

$$p = P(z \leq -1.32 \text{ or } z \geq 1.32) = 2P(z \geq 1.32) = 2(.0934) = .1868$$

EXAMPLE 8.5 Refer to Example 8.1. Compute the observed significance level for the test and interpret its value.

Solution Example 8.1 presented a test of the hypothesis

$$H_0: \mu = 2{,}500$$

against the alternative hypothesis

$$H_a: \mu < 2{,}500$$

The computed value of the test statistic was $z = -1.81$. For this lower-tailed test, the observed significance level is

$$p = P(z \leq -1.81) = .5 - .4649 = .0351$$

This p-value implies that we would reject the null hypothesis for any value of α greater than $.0351$. In Example 8.1, we preselected the significance level $\alpha = .05$, and thus rejected the null hypothesis. However, at a preselected significance level of $\alpha = .01$, the null hypothesis would not be rejected, based on this sample.

EXAMPLE 8.6 Refer to Example 8.3. Compute the observed significance level for the test and interpret its value.

Solution In Example 8.3, we conducted a two-tailed test of the hypothesis

$$H_0: \mu = 16$$

against the alternative hypothesis

$$H_a: \mu \neq 16$$

The value of the test statistic computed from the observed sample was $z = -.95$. Values of the test statistic that would be even more contradictory to H_0 than the computed value are those less than $-.95$ **or** greater than $.95$. Thus, the observed significance level for this test is

$$p = P(z < -.95 \text{ or } z > .95) = 2P(z \geq .95) = 2(.1711) = .3422$$

This large p-value does not cast doubt on the null hypothesis; we would be able to reject H_0 only for the preselected values of α greater than $.3422$.

EXERCISES

8.3 Refer to Exercise 8.1. Compute the observed significance level for the test of hypothesis and interpret its value.

8.4 Refer to Exercise 8.2. Report the *p*-value for the test and give an interpretation.

8.4 Small-Sample Test of Hypothesis About a Population Mean

EXAMPLE 8.7 Refer to Example 7.5. Can we conclude (at the $\alpha = .05$ level) that the average cost of a semi-private room in California hospitals is less than the nationwide average?

Solution It is desired to make an inference about the value of μ, the true mean cost of a semi-private room in all California hospitals. In particular, we wish to test

H_0: $\mu = 350$ (i.e., the average cost in California equals the nationwide average)

against

H_a: $\mu < 350$ (i.e., the average cost in California is less than the nationwide average)

Since the sample size is $n = 20$, we cannot use the large-sample z statistic; thus, we must make the assumption that the distribution of semi-private room costs in California hospitals is approximately normal. The test will then be based on a t distribution with $n - 1 = 19$ degrees of freedom. The rejection region consists of all values of t such that

or $\quad t < -t_\alpha$
or $\quad t < -t_{.05}$
$\quad\quad t < -1.729$

where t is based on 19 df.

The test statistic is

$$t = \frac{\bar{x} - \mu_0}{s/\sqrt{n}} = \frac{336 - 350}{23/\sqrt{20}} = -2.72$$

This value of the test statistic falls within the rejection region. We thus reject H_0 and conclude that the average daily cost of a semi-private room in California hospitals is significantly less than the nationwide average of $350.

EXAMPLE 8.8 A trans-oceanic airline conducted a study to determine whether the mean weight of baggage checked by a passenger on its Miami to London flight differs significantly from 45 pounds. A random sample of 25 passengers was selected, and the weight of each passenger's checked baggage was recorded. The following results were obtained: $\bar{x} = 43.5$ pounds, $s = 6$ pounds. If the airline is willing to risk a Type I error with probability $\alpha = .05$, what should they conclude from this study?

Solution The airline wishes to perform a test of

$$H_0: \mu = 45$$

against

$$H_a: \mu \neq 45$$

where μ is the true mean weight of checked baggage of all passengers on the Miami to London flight.

The test will be based on a t statistic, since $n < 30$, and the large-sample z statistic is inappropriate. We must make the assumption that the baggage weights on Miami to London flights have an approximate normal distribution. Then, for $\alpha = .05$, the rejection region consists of the following sets of t values:

$$t < -t_{\alpha/2} \quad \text{or} \quad t > t_{\alpha/2}$$
i.e., $\quad t < -t_{.025} \quad \text{or} \quad t > t_{.025}$
i.e., $\quad t < -2.064 \quad \text{or} \quad t > 2.064$

where t is based on $n - 1 = 24$ degrees of freedom.

The value of the test statistic is

$$t = \frac{\bar{x} - \mu_0}{s/\sqrt{n}} = \frac{43.5 - 45}{6/\sqrt{25}} = -1.25$$

The computed value of the test statistic does not fall within the rejection region. There is insufficient evidence to conclude that the mean weight of checked baggage of passengers going from Miami to London is significantly different from 45 pounds.

EXERCISES

8.5 Refer to Exercise 7.6. Is there sufficient evidence (at $\alpha = .05$) to conclude that the true mean monthly rental rate for two-bedroom, unfurnished apartments in the area exceeds $500? State any assumptions required for the validity of this hypothesis test.

8.6 A random sample of 22 new homes built last year in a particular state was selected, and the number of square feet of heated floor space was recorded for each. The sample mean and standard deviation were computed as follows: $x = 1,712$; $s = 250$.

a. The average number of square feet of heated floor space for new homes built nationwide is 1,800. Is there sufficient evidence to conclude that new homes in this state have a mean square footage that differs significantly from the nationwide average? Use $\alpha = .05$.

b. State any assumptions required for the validity of the procedure used in part a.

8.5 Large-Sample Test of Hypothesis About a Binomial Probability

EXAMPLE 8.9 Refer to Example 7.8. Is there evidence (at $\alpha = .05$) that the proportion of customers who place additional orders has decreased from previous years?

Solution The mail-order company is interested in a test of hypothesis

$$H_0: p = .10 \text{ (i.e., the proportion has not changed)}$$

against

$$H_a: p < .10 \text{ (i.e., the proportion has decreased)}$$

where p is the proportion of recent customers who place an additional order within six months of the original order.

The test will be based on a z statistic, since the sample size $n = 1,000$ is large enough so that the sampling distribution of $\hat{p}$ is approximately normal. For $\alpha = .05$, the rejection region consists of z values such that

$$z < -z_\alpha$$
or $\quad z < -z_{.05}$
or $\quad z < -1.645$

The test statistic is

$$z = \frac{\hat{p} - p_0}{\sqrt{pq/n}} \approx \frac{\hat{p} - p_0}{\sqrt{p_0 q_0/n}}$$

In our example,

$$\hat{p} = \frac{80}{1,000} = .08, \hat{q} = 1 - \hat{p} = .92, n = 1,000, p_0 = .10$$

Thus,

$$z \approx \frac{.08 - .10}{\sqrt{(.10)(.90)/1,000}} = -2.11$$

Since the computed value of z falls within the rejection region, we reject H_0 and conclude that the proportion of recent customers who place additional orders within six months of the original order is significantly less than .10.

EXERCISES

8.7 A local dairy is considering the possibility of ceasing home deliveries in an effort to minimize cost increases for its products. In a survey of 100 randomly selected customers who currently receive home delivery, 64 customers indicated that they were in favor of this economy measure. The dairy will cease home deliveries if the results of the survey indicate (at $\alpha = .01$) that at least 60% of all home delivery customers favor the proposal. What decision should the dairy make?

8.8 Refer to Example 7.7. A study of last year's returns showed that 10% of all returns contained at least one error. Is there sample evidence (at $\alpha = .01$) that the proportion of this year's returns with at least one error has increased?

8.6 Calculating Type II Error Probabilities: More About β (Optional)

EXAMPLE 8.10 A manufacturing process is tested to determine whether a new machine produces nails with a mean length of 1.2 inches. The test set-up is as follows:

H_0: $\mu = 1.2$
H_a: $\mu \neq 1.2$ (i.e., $\mu < 1.2$ or $\mu > 1.2$)

Test statistic: $z = \dfrac{\bar{x} - 1.2}{\sigma_{\bar{x}}}$

Rejection region: For $\alpha = .05$, $z < -1.96$ or $z > 1.96$
For $\alpha = .01$, $z < -2.575$ or $z > 2.575$

Note that two rejection regions have been specified corresponding to values of $\alpha = .05$ and $\alpha = .01$, respectively. Assume that $n = 100$ and $s = .5$.

a. Suppose that the nails have a mean length of 1.1 inches, i.e., $\mu = 1.1$. Calculate the values of β corresponding to the two rejection regions. Discuss the relationship between the values of α and β.

b. Calculate the power of the test for each of the rejection regions when $\mu = 1.1$.

Solution   a. We first consider the rejection region corresponding to $\alpha = .05$. The first step is to calculate the border values of $\bar{x}$ corresponding to the two-tailed rejection region, $z < -1.96$ or $z > 1.96$:

$$\bar{x}_{0,L} = \mu_0 - 1.96\sigma_{\bar{x}} \approx \mu_0 - 1.96s/\sqrt{n}$$
$$= 1.2 - 1.96(.5/10) = 1.102$$

$$\bar{x}_{0,U} = \mu_0 + 1.96\sigma_{\bar{x}} \approx \mu_0 + 1.96s/\sqrt{n}$$
$$= 1.2 + 1.96(.5/10) = 1.298$$

These border values are shown in the accompanying figure, part a. Next, we convert these values to z values in the alternative distribution with $\mu_a = 1.1$.

$$z_L = \frac{\bar{x}_{0,L} - \mu_a}{\sigma_{\bar{x}}} \approx \frac{1.102 - 1.1}{.05} = .04$$

$$z_U = \frac{\bar{x}_{0,U} - \mu_a}{\sigma_{\bar{x}}} \approx \frac{1.298 - 1.1}{.05} = 3.96$$

These z values are shown in the figure, part b, and you can see that the acceptance (or nonrejection) region is the area between them. Using Table IV of the Appendix, we find that the area between $z = 0$ and $z = .04$ is .0160, and the area between $z = 0$ and $z = 3.96$ is (approximately) .5 (since $z = 3.96$ is off the scale of Table IV). Then the area between $z = .04$ and $z = 3.96$ is, approximately,

$$\beta = .5 - .0160 = .484$$

Thus, the test with $\alpha = .05$ will lead to a Type II error about 48% of the time when the nail mean length is .1 inch less than the expected mean length.

For the rejection region corresponding to $\alpha = .01$, $z < -2.575$ or $z > 2.575$, we find:

$$\bar{x}_{0,L} = 1.2 - 2.575(.5/10) = 1.0712$$

$$\bar{x}_{0,U} = 1.2 + 2.575(.5/10) = 1.3288$$

These border values of the rejection region are shown in the figure, part c.

Converting these to z values in the alternative distribution with $\mu_a = 1.1$, we find $z_L = .58$ and $z_U = 4.58$. The area between these values is, approximately,

$$\beta = .2190 + .5 = .719$$

Thus, the chance that the test procedure with $\alpha = .01$ will lead to an incorrect acceptance of H_0 is about 72%.

Note that the value of β increases from .4840 to .7190 when we decrease the value of α from .05 to .01. This is a general property of the relationship between α and β; as α is decreased (increased), β is increased (decreased).

b. The power is defined to be the probability of (correctly) rejecting the null hypothesis when the alternative is true. When $\mu = 1.1$ and $\alpha = .05$, we find

$$\text{Power} = 1 - \beta = 1 - .4840 = .5160$$

When $\mu = 1.1$ and $\alpha = .01$, we find

$$\text{Power} = 1 - \beta = 1 - .7190 = .2810$$

The powers are shown in the figure, parts **b** and **c**, respectively. You can see that the power of the test is decreased as the level of α is decreased. This means that as the probability of incorrectly rejecting the null hypothesis is decreased, the probability of correctly accepting the null hypothesis for a given alternative is also decreased. Thus, the value of α must be selected carefully, with the realization that a test is made less powerful to detect departures from the null hypothesis when the value of α is decreased.

(a) $\mu = 1.2$ (H_0)
Two rejection regions
$\alpha = .05$ and $\alpha = .01$

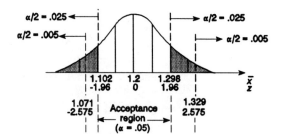

(b) $\mu = 1.1$ (H_a)
β for $\alpha = .05$
rejection region

(c) $\mu = 1.1$ (H_a)
β for $\alpha = .01$
rejection region

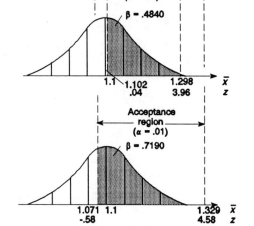

EXERCISE

8.9 Refer to Example 8.10. Suppose the nails have a mean length of 1.35 inches. Calculate β for each of the rejection regions used in the example.

8.7 Inferences about a Population Variance (Optional)

EXAMPLE 8.11 Use the table of critical values of the χ^2 distribution to find the following values of χ^2_α:

a. $\chi^2_{.05}$ with 8 degrees of freedom

b. $\chi^2_{.99}$ with 15 degrees of freedom

Solution a. We first observe that the table of critical values of the χ^2 distribution in your text (Table VII) gives values χ^2_α such that $P(\chi^2 > \chi^2_\alpha) = \alpha$. Now, at the intersection of the column labeled $\chi^2_{.05}$ and the row corresponding to 8 degrees of freedom, we find the entry 15.5073; thus, $P(\chi^2 > 15.5073) = .05$ when df = 8 (see figure):

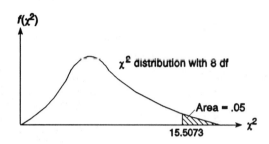

b. At the intersection of the $\chi^2_{.99}$ column and the row corresponding to 15 degrees of freedom, we locate the entry 5.22935. Thus, $P(\chi^2 > 5.22935) = .99$, and hence $P(\chi^2 < 5.22935) = .01$ when df = 15.

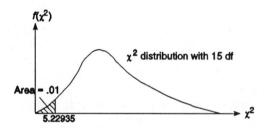

EXAMPLE 8.12 In many states, it is now legal for a pharmacist to substitute a generically equivalent drug for what is prescribed by a physician; this often results in a lower cost for the consumer. However, it is often argued that for the non-namebrand substitutes, the quality control measures are not as strict as for the more popular (and more expensive) brands.

For one drug designed to regulate the heart rate, a random sample of 20 capsules of a generic substitute was selected and the potency (in percent) was measured for each. The sample standard deviation for the measurements was computed, resulting in

$$s = 2.3$$

Construct a 99% confidence interval for the standard deviation of potency for all such capsules of this substitute drug.

Solution	The general form of a confidence interval for a population variance, σ^2, is

$$\frac{(n-1)s^2}{\chi^2_{\alpha/2}} < \sigma^2 < \frac{(n-1)s^2}{\chi^2_{1-\alpha/2}}$$

where the distribution of χ^2 is based on $(n-1)$ degrees of freedom.

In our example, df $= (20 - 1) = 19$,

$$\chi^2_{\alpha/2} = \chi^2_{.005} = 38.5822$$

and

$$\chi^2_{1-\alpha/2} = \chi^2_{.995} = 6.84398$$

Substitution into the general formula yields

$$\frac{19(2.3)^2}{38.5822} < \sigma^2 < \frac{19(2.3)^2}{6.84398}$$

or

$$2.61 < \sigma^2 < 14.69$$

We can be 99% confident that the variance in potency for all capsules of this generic drug is between 2.61 and 14.69; thus, the standard deviation is believed to lie within the interval from $\sqrt{2.61}$ to $\sqrt{14.69}$, or from 1.62 to 3.83 percent.

Note that this confidence interval procedure requires the assumption that the potency measurements have a distribution which is approximately normal.

EXAMPLE 8.13	Refer to Example 8.12. It is known that for the popular brandname equivalent of this drug, the standard deviation of potency measurements is 1.4 percent. Is there evidence to conclude (at $\alpha = .05$) that the standard deviation of potency for the generic substitute exceeds the standard deviation of potency for the brandname equivalent?

Solution	It is required to perform a test of

$$H_0: \sigma^2 = (1.4)^2 = 1.96$$

against

$$H_a: \sigma^2 > 1.96$$

where σ^2 is the variation of potency measurements for all capsules of the generic substitute drug.

At $\alpha = .05$, the rejection region consists of all values of the test statistic χ^2 such that

$$\chi^2 > \chi^2_{.05} = 30.1435$$

where the distribution of χ^2 has $(n - 1) = 19$ degrees of freedom.

$$\chi^2 = \frac{(n-1)s^2}{\sigma_0^2} = \frac{19(2.3)^2}{1.96} = 51.28$$

This value lies within the rejection region; we thus reject H_0 and conclude that the variance (and thus the standard deviation) of potency measurements for the generic substitute significantly exceeds that for the equivalent brandname drug.

This test procedure requires the assumption made in Example 8.12, i.e., that the potency measurements of the generically equivalent substitute drug are approximately normally distributed.

EXERCISES

8.10 Use the table of critical values of the χ^2 distribution to find the following values of χ^2_α:

a. $\chi^2_{.01}$ with 14 degrees of freedom.

b. $\chi^2_{.95}$ with 18 degrees of freedom.

8.11 Find the value of c which makes each of the following statements true:

a. $P(\chi^2 > c) = .995$, where χ^2 has 7 degrees of freedom.

b. $P(\chi^2 < c) = .025$, where χ^2 has 13 degrees of freedom.

8.12 A vending machine which dispenses cold beverages must operate so that the variance of the amount of fill is very small. A random sample of the amounts of fill for 25 drinks produced a standard deviation of .33 ounce. Construct a 90% confidence interval for the true variance of the amount of fill for all drinks dispensed by the machine.

8.13 Refer to Exercise 8.12. If the standard deviation of the amount of fill is significantly higher than .30 ounce, the machine is readjusted. Does the sample evidence in Exercise 8.12 indicate (at $\alpha = .05$) that the machine should be shut down and readjusted?

CHAPTER NINE

Inferences Based on Two Samples: Estimation and Tests of Hypotheses

Summary

This chapter presented techniques for making inferences about the difference between population parameters, based on the information contained in two samples.

Large-sample inferences about the difference ($\mu_1 - \mu_2$) between population means, or the difference ($p_1 - p_2$) between binomial proportions, require minimal assumptions about the sampled populations and use a **two-sample z** statistic. Small-sample inferences about ($\mu_1 - \mu_2$) may be based on an independent samples design (which requires the assumptions of normality and equal population variances) and a **two-sample t** statistic. Alternatively, it is often advantageous to employ a **paired difference design** (to eliminate the effect of variability due to the dimension(s) on which the observations are paired) and a single-sample t statistic to analyze the differences. An **F test** may be used to compare two population variances, σ_1^2 and σ_2^2.

The chapter demonstrated the calculation of the sample size required to estimate ($\mu_1 - \mu_2$) or ($p_1 - p_2$) with a specified degree of accuracy.

9.1 Large-Sample Inferences about the Difference Between Two Population Means: Independent Sampling

EXAMPLE 9.1   A major electrical corporation has recently improved its mercury vapor lamp, which is widely used in street lights. They now claim that the improved bulb has a longer mean lifelength than the leading competitor's mercury vapor bulb. In order to test this claim, a local power company conducted a test with independent random samples of 100 of the improved bulbs and 150 bulbs manufactured by the leading competitor. Each bulb was installed in a city street light and the lifelength for each of the bulbs was recorded, with the following results:

	Improved Bulb	Leading Competitor's Bulb
$\bar{x}$	2,197 hours	2,134 hours
s	123 hours	99 hours
n	100	150

Is this sufficient evidence (at $\alpha = .05$) to substantiate the claim that the improved bulb lasts longer, on the average, than the leading competitor's bulb?

Solution The elements of the relevant hypothesis test are as follows:

$$H_0: (\mu_1 - \mu_2) = 0 \quad (\text{i.e., } \mu_1 = \mu_2)$$
$$H_a: (\mu_1 - \mu_2) > 0 \quad (\text{i.e., } \mu_1 > \mu_2)$$

where μ_1 and μ_2 are the true average lifelengths of the improved bulb and the leading competitor's bulb, respectively.

The sample sizes $n_1 = 100$ and $n_2 = 150$ are sufficient to permit use of a large-sample procedure based on a z statistic. Thus, for a significance level of $\alpha = .05$, we will reject the null hypothesis H_0 if

$$z > z_\alpha, \text{ or } z > z_{.05}, \text{ or } z > 1.645$$

The test statistic is

$$z = \frac{(\bar{x}_1 - \bar{x}_2) - 0}{\sqrt{\frac{\sigma_1^2}{n_1} + \frac{\sigma_2^2}{n_2}}} \approx \frac{(\bar{x}_1 - \bar{x}_2) - 0}{\sqrt{\frac{s_1^2}{n_1} + \frac{s_2^2}{n_2}}}$$

(The sample sizes are sufficient so that σ_1^2 and σ_2^2 may be well approximated by s_1^2 and s_2^2.) For our example, the computed value of z is:

$$z = \frac{2{,}197 - 2{,}134}{\sqrt{\frac{(123)^2}{100} + \frac{(99)^2}{150}}} = 4.28$$

The value of the test statistic falls within the rejection region. We thus reject H_0 and conclude that the average lifelength of the improved bulb is significantly greater than that of the leading competitor's bulb.

EXAMPLE 9.2 Independent random samples of records of new car sales for which a trade-in was accepted were obtained from the local General Motors dealership and from the Ford dealership across the street. The following data were tabulated on the mileage of the trade-in car:

	General Motors	Ford
$\bar{x}$	67,250	58,989
s	22,010	12,308
n	59	45

Construct a 90% confidence interval for the difference between the mean mileages on trade-in cars accepted by the General Motors and Ford dealerships.

Solution The sample sizes are sufficient to employ the large-sample confidence interval procedure for $(\mu_1 - \mu_2)$, where μ_1 and μ_2 denote the mean mileages on trade-in cars accepted by the General Motors and Ford dealerships, respectively. The confidence interval is given by:

$$(\bar{x}_1 - \bar{x}_2) \pm z_{\alpha/2} \sqrt{\frac{\sigma_1^2}{n_1} + \frac{\sigma_2^2}{n_2}} \approx (\bar{x}_1 - \bar{x}_2) \pm z_{\alpha/2} \sqrt{\frac{s_1^2}{n_1} + \frac{s_2^2}{n_2}}$$

$$= (67{,}250 - 58{,}989) \pm 1.645 \sqrt{\frac{(22{,}010)^2}{59} + \frac{(12{,}308)^2}{45}}$$

$$= 8{,}261 \pm 5{,}597, \text{ or } (2{,}664, 13{,}858)$$

We can be 90% confident that the mean mileage on trade-ins accepted by the General Motors dealership is between 2,664 and 13,858 miles higher than the mean mileage on trade-ins accepted by the Ford dealership.

EXAMPLE 9.3 Under what conditions will the sampling distribution of $(\bar{x}_1 - \bar{x}_2)$ be approximately normal?

Solution The Central Limit Theorem guarantees that the sampling distribution of $(\bar{x}_1 - \bar{x}_2)$ will be approximately normal for sufficiently large values of n_1 and n_2, say, $n_1 \geq 30$ and $n_2 \geq 30$.

Further, it can be shown that the mean and variance of the sampling distribution of $(\bar{x}_1 - \bar{x}_2)$ are, respectively:

$$\mu_{(\bar{x}_1 - \bar{x}_2)} = E(\bar{x}_1 - \bar{x}_2) = \mu_1 - \mu_2$$

and

$$\sigma^2_{(\bar{x}_1 - \bar{x}_2)} = \frac{\sigma_1^2}{n_1} + \frac{\sigma_2^2}{n_2}$$

assuming the two samples are randomly selected in an independent manner from the two populations.

EXERCISES

9.1 For many physical science majors, the choice of an area of concentration is often related to the job market in the area. Independent random samples of jobs for recent graduating seniors in physics and chemistry produced the following information regarding starting monthly salaries:

	Physics	Chemistry
$\bar{x}$	$2,421	$2,376
s	$72	$63
n	41	38

Is there sufficient evidence (at $\alpha = .05$) to conclude that the mean starting monthly salaries for physics and chemistry majors are significantly different?

9.2 Refer to Exercise 9.1. Construct a 95% confidence interval for the difference in mean starting monthly salaries between physics and chemistry majors. Interpret the confidence interval.

9.2 Small-Sample Inferences About the Difference Between Two Population Means: Independent Sampling

EXAMPLE 9.4 An educational researcher wishes to determine if there is a difference in the development of musical ability between boys and girls. Six second-grade boys and four second-grade girls were given a 20-point test to evaluate their musical talents. The means and standard deviations of the scores are shown below:

	Boys	Girls
$\bar{x}$	14.82	16.80
s	1.12	1.46

Is there evidence (at $\alpha = .05$) that second-grade girls have greater musical ability, as measured by this test, than second-grade boys?

Solution Since the sample sizes are small, the test must be based on a two-sample t statistic, and the following assumptions are required:

(i) The populations of test scores must be approximately normally distributed, for both second-grade boys and second-grade girls.

(ii) The variances of the two populations must be equal.

(iii) The two samples are randomly selected in an independent manner from the two populations.

The researcher is then interested in a test of the hypothesis

$$H_0: (\mu_1 - \mu_2) = 0 \text{ (i.e., } \mu_1 = \mu_2)$$

against

$$H_a: (\mu_1 - \mu_2) < 0 \text{ (i.e., } \mu_1 < \mu_2)$$

where μ_1 and μ_2 are the population mean scores on the test, for all second-grade boys and girls, respectively. Note that the alternative hypothesis of interest ($H_a: \mu_1 - \mu_2 < 0$) states that the mean test score for boys is lower than that for girls.

At $\alpha = .05$, we will reject H_0 if $t < -t_{.05}$, where t has $n_1 + n_2 - 2 = 6 + 4 - 2 = 8$ degrees of freedom. From the table of critical values for the t distribution, we see that for df = 8, $t_{.05} = 1.860$. The rejection region thus consists of all values of t such that $t < -1.860$.

It is now required to compute s_p^2, the pooled estimate of the common population variance. We obtain:

$$s_p^2 = \frac{(n_1 - 1)s_1^2 + (n_2 - 1)s_2^2}{n_1 + n_2 - 2} = \frac{(6 - 1)(1.12)^2 + (4 - 1)(1.46)^2}{6 + 4 - 2}$$

$$= \frac{5(1.2544) + 3(2.1316)}{8} = 1.58335$$

Then the value of the test statistic is

$$t = \frac{(\bar{x}_1 - \bar{x}_2) - 0}{\sqrt{s_p^2 \left[\frac{1}{n_1} + \frac{1}{n_2}\right]}} = \frac{14.82 - 16.80}{\sqrt{1.58335 \left[\frac{1}{6} + \frac{1}{4}\right]}} = \frac{-1.98}{.8122} = -2.44$$

Since the computed value of the test statistic falls within the rejection region, we reject H_0 and conclude that the mean test score for second-grade girls is significantly higher than that for boys.

EXAMPLE 9.5 Independent random samples of two well-advertised brands of house paint produced the following results on coverage (in square feet of surface) per gallon:

	Paint A	Paint B
$\bar{x}$	615	570
s	42	31
n	10	10

Estimate the difference in mean coverage for the two paints with a 90% confidence interval.

Solution The following assumptions must be made in order to employ the small-sample confidence interval procedure:

(i) The populations of coverage per gallon values must be normally distributed for both Paint A and Paint B.

(ii) The population variances must be equal for the two populations.

(iii) The samples must be obtained randomly and in an independent manner from the two populations.

Let us define μ_1 and μ_2 to be the mean coverage (in square feet of surface) per gallon for Paint A and Paint B, respectively. We wish to obtain a 90% confidence interval for $(\mu_1 - \mu_2)$, the difference in mean coverage for the two paints.

The pooled estimate of the common population variance is

$$s_p^2 = \frac{(n_1 - 1)s_1^2 + (n_2 - 1)s_2^2}{n_1 + n_2 - 2} = \frac{9(42)^2 + 9(31)^2}{18} = 1,362.5$$

Now, the 90% confidence interval is based on the critical value of $t_{\alpha/2} = t_{.05} = 1.734$, where t has $n_1 + n_2 - 2 = 18$ degrees of freedom. The desired confidence interval is then:

$$(\bar{x}_1 - \bar{x}_2) \pm t_{.05} \sqrt{s_p^2 \left(\frac{1}{n_1} + \frac{1}{n_2}\right)}$$

$$= (615 - 570) \pm 1.734 \sqrt{(1,362.5)\left(\frac{1}{10} + \frac{1}{10}\right)}$$

$$= 45 \pm 28.6, \text{ or } (16.4, 73.6)$$

We can be 90% confident that the mean coverage obtained per gallon of Paint A exceeds the mean coverage per gallon of Paint B by between 16.4 and 73.6 square feet.

EXERCISES

9.3 Automotive researchers have recently been interested in comparing the amount of air pollution for rotary engine and piston engine cars. A test designed to measure the amount of pollution (in milligrams per cubic yard of exhaust) produced the following results:

	Rotary Engine	Piston Engine
$\bar{x}$	70.8	79.0
s	29.0	23.0
n	8	4

a. Test the hypothesis (at $\alpha = .01$) that the mean pollution emitted by rotary engine cars is significantly lower than the mean pollution emitted by piston engine cars.

b. What assumptions are required for the validity of the procedure used in part **a**?

9.4 Refer to Exercise 9.3. Construct a 95% confidence interval for the difference in the mean amounts of pollution emitted by rotary engine and piston engine cars. Interpret the confidence interval.

9.3 Inferences About the Difference Between Two Population Means: Paired Difference Experiments

EXAMPLE 9.6 Due to a possible energy shortage, the federal government has been funding researchers to investigate and develop alternative sources of fuel. A new product, methanol, is currently being tested in California. It costs much less than regular gasoline and, if judged successful, will require fewer natural oil resources than gasoline does. A test was run on 5 cars of different sizes to compare the mileage per gallon obtained using methanol with the mileage per gallon obtained with a standard brand of gasoline. The following results were obtained:

	Mileage Per Gallon		
Car	Methanol	Regular Gasoline	Difference x_D
1	14.0	12.3	1.7
2	20.0	17.8	2.2
3	21.4	20.9	.5
4	27.1	25.7	1.4
5	35.6	33.0	2.6

Can we conclude (at $\alpha = .01$) that the mean mileage per gallon of methanol is significantly greater than the mean mileage per gallon obtained with regular gasoline?

Solution

Based on this paired difference experiment, we wish to test the hypothesis

$$H_0: \mu_D = 0 \quad (\text{i.e., } \mu_1 - \mu_2 = 0)$$
$$H_a: \mu_D > 0 \quad (\text{i.e., } \mu_1 - \mu_2 > 0)$$

where μ_1 and μ_2 are the population mean mileage per gallon ratings obtained by cars using methanol and regular gasoline, respectively, and $\mu_D = \mu_1 - \mu_2$.

The assumptions required are:

(i) The population of differences in mean mileage per gallon ratings is normally distributed.

(ii) The sample differences are randomly selected from the population of differences.

At $\alpha = .01$, we will reject the null hypothesis for all values of t such that $t > t_{.01} = 3.747$, where t has $n_D - 1 = 4$ degrees of freedom.

It is necessary to compute $\bar{x}_D$ and s_D from the sample differences:

$$\bar{x}_D = \frac{1.7 + 2.2 + .5 + 1.4 + 2.6}{5} = 1.68$$

and

$$s_D = \sqrt{\frac{\sum x_D^2 - \frac{(\sum x_D)^2}{n_D}}{n_D - 1}}$$

$$= \sqrt{\frac{(1.7)^2 + (2.2)^2 + (.5)^2 + (1.4)^2 + (2.6)^2 - \frac{(8.4)^2}{5}}{4}}$$

$$= \sqrt{\frac{16.7 - 14.112}{4}} = \sqrt{.617} = .8044$$

Then the test statistic is

$$t = \frac{\bar{x}_D - 0}{s_D/\sqrt{n_D}} = \frac{1.68}{.8044/\sqrt{5}} = 4.67$$

Since the value of the test statistic is within the rejection region, we reject the null hypothesis and conclude that the mean mileage per gallon rating for cars using methanol is significantly greater (at $\alpha = .01$) than the mean mileage rating for cars using regular gasoline.

EXAMPLE 9.7 In order to compare sales for two local fast-food chains, a marketing researcher recorded the number of hamburgers sold during the noon hour for each day during a randomly selected week. The following results were obtained:

	Chain A	B	Difference
Monday	137	97	40
Tuesday	98	63	35
Wednesday	69	59	10
Thursday	72	48	24
Friday	83	62	21
Saturday	142	101	41
Sunday	126	105	21
			$\bar{x}_D = 27.43$
			$s_D = 11.53$

Construct a 95% confidence interval for $(\mu_1 - \mu_2)$, where μ_1 and μ_2 are the mean noon-hour sales (number of hamburgers) for fast-food chains A and B, respectively.

Solution The confidence interval procedure for the paired difference experiment requires the following assumptions:

(i) The population of differences in noon-hour sales is normally distributed.

(ii) The sample differences are randomly selected from the population of differences.

The general form of a 95% confidence interval for $(\mu_1 - \mu_2)$ is

$$\bar{x}_D \pm t_{.025}\left(s_D/\sqrt{n_D}\right)$$

where t is based on $n_D - 1 = 6$ degrees of freedom. For this example, we have

$$27.43 \pm 2.447\left(11.53/\sqrt{7}\right) = 27.43 \pm 10.66, \text{ or } (16.77, 38.09)$$

The marketing researcher can be 95% confident that the mean number of hamburgers sold during the noon hour in the Chain A store is between 16.77 and 38.09 higher than the mean noon-hour sales in the Chain B store.

EXERCISES

9.5 A local health spa advertises that its clients lose an average of 10 pounds during their four-week weight reduction program. An investigation by the Better Business Bureau produced the following data on four people enrolled in the program:

Individual	Weight Before	Weight After
1	203	195
2	144	132
3	157	155
4	128	120

a. Is there evidence (at $\alpha = .01$) to refute the claim of the health spa?
(HINT: Perform a test of the hypothesis H_0: $\mu_D = 10$ against H_a: $\mu_D < 10$.)

b. What assumptions are required for the validity of the procedure used in part **a**?

9.6 Due to many complaints about the high costs of textbooks at the campus bookstore, a student opened an off-campus bookstore. He claims that the average price of textbooks at his store is less than the average price charged at the campus bookstore. A comparison of prices for 10 high-volume textbooks revealed the following:

Textbook	Campus Bookstore Price	Off-Campus Bookstore Price
1	$42.95	$40.95
2	38.00	37.00
3	35.50	36.00
4	41.00	39.95
5	46.00	44.95
6	38.00	37.50
7	43.50	42.00
8	34.95	33.50
9	37.95	35.75
10	42.00	40.50

a. Construct a 95% confidence interval for the mean difference in textbook prices charged by the two bookstores.

b. State the assumptions upon which the confidence interval procedure in part **a** is based.

9.7 A company pays its salesmen on a commission basis and trains them to sell the product either on a door-to-door basis or by telephone contact. They wish to perform an experiment to compare the effectiveness of these two techniques; the following designs have been proposed:

Design A: Select ten salesmen, five of whom will be randomly assigned to sell door-to-door for one week and the remaining five will sell by telephone contact for one week. Compare the mean sales for the two methods.

Design B: Select five salesmen, and randomly assign the method of sale for the first week; the other method will be used the following week. Compare the results for the two methods.

a. Identify the two types of designs and state the assumptions required for making inferences from each.

b. Which design do you think is preferable in this situation? Why?

9.4 Inferences About the Difference Between Population Proportions: Independent Binomial Experiments

EXAMPLE 9.8 In a national political poll of 1,500 voters conducted in January, 87 people expressed strong disapproval of the proposed federal budget for the next fiscal year. An independent poll of 1,500 voters conducted in June indicated that 112 people strongly disapproved of the proposed budget. Is there evidence (at $\alpha = .01$) that the proportion of voters who strongly disapproved of the federal budget changed in the period from January to June?

Solution We are interested in a test of

$$H_0: (p_1 - p_2) = 0 \text{ (i.e., } p_1 = p_2)$$

against

$$H_a: (p_1 - p_2) \neq 0 \text{ (i.e., } p_1 \neq p_2)$$

where p_1 and p_2 are the proportions of voters in January and June, respectively, who strongly disapproved of the proposed federal budget for the next fiscal year.

The sample sizes are sufficiently large that we may employ a z statistic. For $\alpha = .01$, we will reject H_0 if

$$z < -z_{.005}, \text{ or } z > z_{.005}$$

i.e., if

$$z < -2.58, \text{ or } z > 2.58$$

The following quantities are required for computation of the test statistic:

$$\hat{p}_1 = \frac{\text{Number of voters who strongly disapproved in January}}{\text{Number of voters in January sample}} = \frac{87}{1{,}500} = .0580$$

$$\hat{p}_2 = \frac{\text{Number of voters who strongly disapproved in June}}{\text{Number of voters in June sample}} = \frac{112}{1{,}500} = .0747$$

$$\hat{q}_1 = 1 - \hat{p}_1 = 1 - .0580 = .9420$$

$$\hat{q}_2 = 1 - \hat{p}_2 = 1 - .0747 = .9253$$

Our pooled estimate of the common value of $p_1 = p_2 = p$, to be used in the denominator of the test statistic, is

$$\hat{p} = \frac{87 + 112}{1{,}500 + 1{,}500} = \frac{199}{3{,}000} \approx .0663$$

Then $\hat{q} = 1 - \hat{p} = 1 - .0663 = .9337$, and the test statistic is

$$z = \frac{(\hat{p}_1 - \hat{p}_2) - 0}{\sqrt{\hat{p}\hat{q}\left[\frac{1}{n_1} + \frac{1}{n_2}\right]}} = \frac{(.0580 - .0747) - 0}{\sqrt{(.0663)(.9337)\left[\frac{1}{1{,}500} + \frac{1}{1{,}500}\right]}} = -1.84$$

This value of the test statistic does not lie within the rejection region. There is insufficient evidence (at $\alpha = .01$) to conclude that the proportion of voters who strongly disapproved of the federal budget changed significantly from January to June.

EXAMPLE 9.9 A life insurance salesman wishes to compare the proportion of contacts to married and single individuals which result in sales. An examination of his records for the past three months revealed that he had sold policies to 24 out of 52 married people contacted, and to 19 out of 60 single people contacted. Construct a 95% confidence interval for the difference in the proportions of married and single individuals who purchase policies.

Solution Let p_1 and p_2 be the proportions of married people and single people, respectively, who purchase policies from the salesman. A 95% confidence interval for $(p_1 - p_2)$ is given by

$$(\hat{p}_1 - \hat{p}_2) \pm z_{.025}\sqrt{\frac{\hat{p}_1\hat{q}_1}{n_1} + \frac{\hat{p}_2\hat{q}_2}{n_2}}$$

where

$$\hat{p}_1 = \frac{24}{52} = .46,\; \hat{p}_2 = \frac{19}{60} = .32,\; \hat{q}_1 = 1 - .46 = .54$$

$\hat{q}_2 = 1 - .32 = .68$, $n_1 = 52$, $n_2 = 60$, and $z_{.025} = 1.96$

Substitution of these values yields:

$$(.46 - .32) \pm 1.96\sqrt{\frac{(.46)(.54)}{52} + \frac{(.32)(.68)}{60}} = .14 \pm .18, \text{ or } (-.04, .32)$$

We estimate that p_2, the proportion of single people who purchase policies, could be larger than p_1, the proportion of married people who purchase policies, by as much as .04, or p_2 could be less than p_1 by as much as .32.

EXERCISES

9.8 A recent survey was taken to compare the housing preferences of senior citizens and citizens under the age of 65. The results are summarized below:

	Senior Citizens	Citizens under age 65
Number of people surveyed	500	400
Number of people in sample who prefer multiple-family housing accommodations	180	120

Is there sufficient evidence (at $\alpha = .01$) to conclude that the proportion of senior citizens who prefer multiple-family housing accommodations is significantly higher than the proportion of citizens under age 65 who prefer such accommodations?

9.9 Refer to Exercise 9.8. Construct a 95% confidence interval for the difference in the proportions of senior citizens and citizens under age 65 who prefer multiple-family housing accommodations.

9.5 Determining the Sample Size

EXAMPLE 9.10 A study is being designed to compare the scores on the Law School Admission Test (LSAT) for history and non-history majors. Past testing has shown that the standard deviation of scores for all people taking the test is approximately 80 points. How many students should be sampled in order to estimate the difference in mean LSAT scores between history and non-history majors, if we want to be 95% confident that the estimate is within 10 points of the true mean difference?

Solution It is required to solve the equation

$$z_{\alpha/2} \sqrt{\frac{\sigma_1^2}{n_1} + \frac{\sigma_2^2}{n_2}} = B$$

where $z_{\alpha/2} = z_{.025} = 1.96$ and $B = 10$. We will assume that $\sigma_1^2 = \sigma_2^2 = 80^2$, where σ_1^2 and σ_2^2 are the population variances of scores for history and non-history majors, respectively. If the two sample sizes are to be equal ($n_1 = n_2 = n$), then

$$1.96 \sqrt{\frac{80^2}{n} + \frac{80^2}{n}} = 10$$

We now solve for n:

$$1.96 \sqrt{\frac{2(80)^2}{n}} = 10 \text{ or } n = 491.7$$

Thus, we should randomly sample 492 history and 492 non-history majors in order to be 95% confident that the estimated difference in mean LSAT scores is correct to within 10 points.

EXAMPLE 9.11 A tobacco company wishes to estimate the difference in the proportions of men and women who smoke cigarettes. How many men and women must be included in their samples if the company wishes to be 90% confident that the estimated difference is within .02 of the true difference in proportions? Past studies indicate that the fraction of smokers among both sexes is approximately .25.

Solution We wish to estimate $(p_1 - p_2)$, where p_1 and p_2 are the proportions of men and women, respectively, who smoke. For the desired confidence, $z_{\alpha/2} = z_{.05} = 1.645$. Then, with $p_1 = p_2 = .25$ and equal sample sizes $n_1 = n_2 = n$, the required sample sizes are found by solving the following equation for n:

$$z_{\alpha/2} \sqrt{\frac{p_1 q_1}{n_1} + \frac{p_2 q_2}{n_2}} = B$$

$$1.645 \sqrt{\frac{(.25)(.75)}{n} + \frac{(.25)(.75)}{n}} = .02$$

or

$$1.645 \sqrt{\frac{2(.25)(.75)}{n}} = .02$$

$$n = 2{,}536.9$$

The tobacco company should sample at least 2,537 men and 2,537 women in order to estimate $(p_1 - p_2)$ with the specified level of reliability.

EXERCISES

9.10 A television station wishes to estimate the difference in the proportions of viewers who watch their 6 P.M. and 11 P.M. newscasts. How many 6 P.M. and 11 P.M. viewers should be sampled in order to be 95% confident that the estimated difference is within .07 of the true difference in proportions? Past surveys indicate that approximately 60% of 6 P.M. viewers and 80% of 11 P.M. viewers watch the station's newscasts.

9.11 A state welfare agency is interested in estimating the difference in mean weekly food costs between families with one child and families with no children. How many one-child and no-child families should be sampled in order to be 90% confident that the estimated difference is within $2.00 of the true difference in mean costs? Previous studies have shown that the standard deviation of weekly food costs is approximately $7.50 for no-child and one-child families.

9.6 Comparing Two Population Variances: Independent Random Samples

EXAMPLE 9.12 Obtain the critical values of the F distribution for the following situations:

a. $F_{.01}$, numerator df = 8, denominator df = 6

b. $F_{.01}$, numerator df = 5, denominator df = 120

c. $F_{.05}$, numerator df = 12, denominator df = 24

Solution

a. In the table of percentage points of the F distribution for $\alpha = .01$, we locate the entry at the intersection of the column for 8 numerator degrees of freedom and the row for 6 denominator degrees of freedom: $F_{.01} = 8.10$. Thus, for an F distribution with 8 numerator and 6 denominator degrees of freedom, $P(F > 8.10) = .01$.

b. In the table of percentage points for $\alpha = .01$, we locate the entry at the intersection of the column for 5 numerator degrees of freedom and the row for 120 denominator degrees of freedom: $F_{.01} = 3.17$.

c. In the table of percentage points of the F distribution for $\alpha = .05$, we find $F_{.05} = 2.18$, where F has 12 numerator and 24 denominator degrees of freedom.

EXAMPLE 9.13 Two corporation executives have turned in last month's receipts for business expenses. The following summarizes the information regarding expenses for "business lunches" by the two executives:

	Executive A	Executive B
$\bar{x}$	$38.42	$40.75
s	$5.40	$3.70
n	13	10

In order to perform the t test (Section 9.2) for the difference between mean expenditures for business lunches by the two executives, we must make the assumption of equal population variances. Is such an assumption reasonable, based on these results? Use a significance level of $\alpha = .10$.

Solution We wish to test the hypothesis

$$H_0: \frac{\sigma_1^2}{\sigma_2^2} = 1 \quad (\text{i.e., } \sigma_1^2 = \sigma_2^2)$$

against

$$H_a: \frac{\sigma_1^2}{\sigma_2^2} \neq 1 \quad (\text{i.e., } \sigma_1^2 \neq \sigma_2^2)$$

where σ_1^2 and σ_2^2 are the population variances of the expenses for business lunches by Executives A and B, respectively.

Inferences Based on Two Samples: Estimation and Tests of Hypotheses

The test requires the following assumptions:

(i) The populations of business lunch expenses are normally distributed for each executive.

(ii) The samples are obtained randomly and independently from the two populations.

The rejection region consists of values of the F statistic for which $F > F_{\alpha/2}$, or $F > F_{.05}$, where F has $n_1 - 1 = 12$ numerator degrees of freedom and $n_2 - 1 = 9$ denominator degrees of freedom. From the table of percentage points for $\alpha = .05$, we obtain the critical value $F_{.05} = 3.07$. We will thus reject H_0 if $F > 3.07$.

The value of the test statistic is

$$F = \frac{\text{Larger sample variance}}{\text{Smaller sample variance}} = \frac{(5.40)^2}{(3.70)^2} = 2.13$$

The computed value of the test statistic does not lie within the rejection region. There is insufficient evidence to conclude that the population variances are significantly different at $\alpha = .10$.

EXAMPLE 9.14 Refer to Example 9.13. Construct a 90% confidence interval for σ_1^2/σ_2^2, the ratio of the population variances of the expenses for business lunches by the two executives.

Solution The general form of a 90% confidence interval for the ratio of two population variances is

$$\left[\frac{s_1^2}{s_2^2}\right]\left[\frac{1}{F_{L,.05}}\right] < \frac{\sigma_1^2}{\sigma_2^2} < \left[\frac{s_1^2}{s_2^2}\right] F_{U,.05}$$

In our example,

$$s_1^2 = (5.40)^2 = 29.16$$
$$s_2^2 = (3.70)^2 = 13.69$$

$F_{L,.05} = 3.07$ (based on $n_1 - 1 = 12$ numerator df and $n_2 - 1 = 9$ denominator df)

and

$F_{U,.05} = 2.80$ (based on $n_2 - 1 = 9$ numerator df and $n_1 - 1 = 12$ denominator df)

Substitution of these values into the general expression yields

$$\left[\frac{29.16}{13.69}\right]\left[\frac{1}{3.07}\right] < \frac{\sigma_1^2}{\sigma_2^2} < \left[\frac{29.16}{13.69}\right](2.80)$$

$$.694 < \frac{\sigma_1^2}{\sigma_2^2} < 5.964$$

EXERCISES

9.12 Obtain the critical values of the F distribution in each of the following situations.

a. $F_{.01}$, numerator df = 3, denominator df = 20

b. $F_{.05}$, numerator df = 30, denominator df = 60

c. $F_{.05}$, numerator df = 24, denominator df = 9

9.13 Refer to Exercise 9.3. One of the assumptions required for the hypothesis test is that the population variances of pollution amounts emitted by rotary engines and piston engines are equal. Do the data provide sufficient evidence (at $\alpha = .02$) to refute the assumption of equal population variances?

9.14 Refer to Exercise 9.13. Construct a 98% confidence interval for the ratio of the population variances of pollution amounts emitted by rotary engines and piston engines.

CHAPTER TEN

Analysis of Variance: Comparing More Than Two Means

Summary

This chapter presented experimental designs useful for comparing more than two means. The **independent sampling design (completely randomized design)** uses p independent random samples to compare the means of p populations. The **randomized block design** is an extension of the paired difference design which uses relatively **homogeneous blocks** of experimental units. Each of p **treatments** is randomly assigned to one experimental unit in each block to compare the treatment means.

The method of analysis for either design involves a comparison of the variation among the treatment means (measured by MST, the mean square for treatments) to the variation among experimental units (measured by MSE, the mean square for error). If the ratio MST/MSE is large, we conclude that the means of at least two of the populations differ.

The analysis of variance procedure can also be used to analyze data from a **two-factor factorial experiment**. Such an analysis allows us to test for **interaction** among factors.

The chapter included examples of the use of the Bonferroni technique to compare all pairs of treatment means.

10.2 The Completely Randomized Design

EXAMPLE 10.1 The nursing director at a privately-owned hospital wishes to compare the weekly mean number of complaints received against the nursing staff during the three daily shifts: First (7 A.M. - 3 P.M.), Second (3 P.M. - 11 P.M.), and Third (11 P.M. - 7 A.M.). Her plan is to select random samples of weekly records from each shift, and record the number of complaints received during the week. What type of experimental design is this?

Solution The nursing director is employing an independent sampling design (completely randomized design). In this example, the means of $p = 3$ populations (numbers of weekly complaints against First, Second, and Third shifts) are to be compared by selecting independent random samples from each of the populations.

EXAMPLE 10.2 Refer to Example 10.1. Set up the test to compare the weekly mean number of complaints against the three shifts. (Data will be provided in the next example.)

Solution We wish to test the null hypothesis that the weekly mean number of complaints is the same for each of the three shifts, against the alternative that at least two of the means are different:

against
$$H_0: \mu_1 = \mu_2 = \mu_3$$
$$H_a: \text{At least two of the means differ}$$

where μ_1, μ_2, and μ_3 are the true mean number of complaints received weekly against the First, Second, and Third shifts, respectively.

The test is based upon the following assumptions:

(i) The distributions of the number of weekly complaints against each of the three shifts are approximately normal.
(ii) The variance of the number of weekly complaints is the same for each of the three shifts.

The test statistic is $F = \text{MST/MSE}$ and the null hypothesis is rejected (at significance level α) if $F > F_\alpha$, where the distribution of F is based on $p - 1 = 3 - 1 = 2$ numerator degrees of freedom and $n - p = n - 3$ denominator degrees of freedom.

EXAMPLE 10.3 Refer to Examples 10.1 and 10.2. The data for the experiment are shown below.

	Number of Weekly Complaints		
	First Shift	Second Shift	Third Shift
	9	8	15
	9	11	14
	11	6	10
	9	8	11
	12	9	10
	—	12	12
Totals	50	54	72

Perform the test to compare the mean number of complaints for the three shifts. Use $\alpha = .05$.

Solution [Note: In all examples in this chapter, we use the calculation formulas for analysis of variance given in Appendix B of the text.]

The following calculations are required:

$$n = n_1 + n_2 + n_3 = 5 + 6 + 6 = 17$$
$$\sum x = T_1 + T_2 + T_3 = 50 + 54 + 72 = 176$$
$$\sum x^2 = 9^2 + 9^2 + 11^2 + \cdots + 11^2 + 10^2 + 12^2 = 1,904$$
$$CM = \frac{(\sum x)^2}{n} = \frac{(176)^2}{17} = 1,822.12$$

Analysis of Variance: Comparing More Than Two Means

$$SS(\text{Total}) = \sum x^2 - CM = 1{,}904 - 1{,}822.12 = 81.88$$

$$SST = \frac{T_1^2}{n_1} + \frac{T_2^2}{n_2} + \frac{T_3^2}{n_3} - CM = \frac{50^2}{6} + \frac{54^2}{6} + \frac{72^2}{6} - 1{,}822.12$$
$$= 1{,}850 - 1{,}822.12 = 27.88$$

$$SSE = SS(\text{Total}) - SST = 81.88 - 27.88 = 54.00$$

$$MST = \frac{SST}{p-1} = \frac{27.88}{2} = 13.94$$

$$MSE = \frac{SSE}{n-p} = \frac{54.00}{17-3} = \frac{54.00}{14} = 3.86$$

Then the computed value of the test statistic is

$$F = \frac{MST}{MSE} = \frac{13.94}{3.86} = 3.61$$

The critical value of F is based on $p - 1 = 2$ numerator df and $n - p = 14$ denominator df. Thus, for $\alpha = .05$, the rejection region consists of values of F such that $F > F_{.05} = 3.74$.

Since the value of the test statistic does not lie within the rejection region, there is insufficient evidence to conclude that a significant difference exists among the weekly mean number of complaints against the different shifts.

EXAMPLE 10.4 Refer to Examples 10.1 - 10.3. Summarize the results of the analysis of variance in an ANOVA table.

Solution The ANOVA summary table for this completely randomized design is as follows:

Source	df	SS	MS	F
Shift	2	27.88	13.94	3.61
Error	14	54.00	3.86	
Total	16	81.88		

EXAMPLE 10.5 Refer to Example 10.3. Construct a 90% confidence interval for μ_1, the weekly mean number of complaints against the nursing staff of the first shift.

Solution The general form of a 90% confidence interval for μ_1 is

$$\bar{x}_1 \pm t_{\alpha/2} \frac{s}{\sqrt{n_1}}$$

where the distribution of t is based on $(n - p)$ degrees of freedom and $s = \sqrt{MSE}$ is the estimate of σ, the common standard deviation of the p populations. In the nursing complaints example, we have

$$\bar{x}_1 = \frac{T_1}{n_1} = \frac{50}{5} = 10$$

$$s = \sqrt{MSE} = \sqrt{3.86} \approx 1.96$$

and $t_{\alpha/2} = t_{.05} = 1.761$ (with $n - p = 17 - 3 = 14$ df). Thus, the required confidence interval is

$$\bar{x}_1 \pm t_{\alpha/2} \frac{s}{\sqrt{n_1}} = 10.0 \pm 1.761 \left[\frac{1.96}{\sqrt{5}}\right] = 10.0 \pm 1.5, \text{ or } (8.5, 11.5)$$

We are 90% confident that the weekly mean number of complaints against the nursing staff of the first shift lies within this interval.

EXAMPLE 10.6 Refer to Example 10.3.

a. Construct a 95% confidence interval for the difference in the weekly mean numbers of complaints against the third and first shifts at the hospital.

b. Interpret the interval.

Solution a. We wish to obtain a 95% confidence interval for $(\mu_3 - \mu_1)$; the general form is

$$(\bar{x}_3 - \bar{x}_1) \pm t_{\alpha/2} s \sqrt{\frac{1}{n_3} + \frac{1}{n_1}}$$

where $s = \sqrt{MSE}$ and t is based on $(n - p)$ degrees of freedom. For the data given in Example 10.3,

$$\bar{x}_3 = \frac{T_3}{n_3} = \frac{72}{6} = 12.0$$

$$\bar{x}_1 = \frac{T_1}{n_1} = \frac{50}{5} = 10.0$$

$$s = \sqrt{MSE} = \sqrt{3.86} \approx 1.96$$

and $t_{\alpha/2} = t_{.025} = 2.145$ (14 df)

Thus, a 95% confidence interval for $(\mu_3 - \mu_1)$ is given by

$$(\bar{x}_3 - \bar{x}_1) \pm t_{.025} s \sqrt{\frac{1}{n_3} + \frac{1}{n_1}} = (12.0 - 10.0) \pm 2.145(1.96)\sqrt{\frac{1}{6} + \frac{1}{5}}$$

$$= 2.0 \pm 2.5, \text{ or } (-.5, 4.5)$$

b. We estimate, with 95% confidence, that μ_1, the weekly mean number of complaints against the first shift, could be larger than μ_3, the weekly mean number of complaints against the third shift, by as much as .5, or it could be less than μ_3 by 4.5. Note that the confidence interval contains zero; this is not surprising since (in Example 10.3) we failed to reject the hypothesis of equality of the weekly mean numbers of complaints against the three shifts.

EXAMPLE 10.7 A durability test was performed on three of the most heavily advertised brands of flashlight batteries. Six batteries of each brand were tested and the time (in hours) was recorded at which a point less than 90% of full power was reached. The results are shown below:

	Brand 1	Brand 2	Brand 3
	59	61	38
	43	63	42
	47	57	38
	45	49	43
	51	48	40
	53	60	46
Totals	298	338	247

a. Is there evidence (at $\alpha = .05$) of a difference in the mean durabilities of the three brands of batteries?

b. Estimate the mean durability of Brand 2 batteries with a 90% confidence interval.

c. Estimate the difference between the mean durabilities of Brands 2 and 3 with a 95% confidence interval.

d. What assumptions are required for the validity of the procedures used in parts a – c?

Solution

a. The elements of the hypothesis test are the following:

H_0: $\mu_1 = \mu_2 = \mu_3$
H_a: At least two of the means are different

where μ_1, μ_2, and μ_3 are the mean durabilities of all batteries of Brands 1, 2 and 3, respectively.

At $\alpha = .05$, the null hypothesis will be rejected for all values of the test statistic F such that $F > F_{.05} = 3.68$, where the distribution of F is based on $p - 1 = 3 - 1 = 2$ numerator degrees of freedom and $n - p = 18 - 3 = 15$ denominator degrees of freedom.

The following computations are required:

$$\sum x = 883$$

$$\sum x^2 = 44,435$$

$$CM = \frac{(\sum x)^2}{n} = \frac{(883)^2}{18} = 43,316.056$$

$$SS(\text{Total}) = \sum x^2 - CM = 44,435 - 43,316.056 = 1,118.944$$

$$\text{SST} = \frac{T_1^2}{n_1} + \frac{T_2^2}{n_2} + \frac{T_3^2}{n_3} - \text{CM}$$

$$= \frac{(298)^2}{6} + \frac{(338)^2}{6} + \frac{(247)^2}{6} - 43{,}315.056 = 693.444$$

$$\text{SSE} = \text{SS(Total)} - \text{SST} = 1{,}118.944 - 693.444 = 425.5$$

$$\text{MST} = \frac{\text{SST}}{p-1} = \frac{693.444}{3-1} = 346.722$$

$$\text{MSE} = \frac{\text{SSE}}{n-p} = \frac{425.5}{18-3} = 28.367$$

Now, the test statistic is

$$F = \frac{\text{MST}}{\text{MSE}} = \frac{346.722}{28.367} = 12.22$$

This value lies within the rejection region; we thus conclude (at $\alpha = .05$) that the mean durabilities for at least two of the brands of batteries differ.

b. A 90% confidence interval for μ_2 is given by

$$\bar{x}_2 \pm t_{.05} \frac{s}{\sqrt{n_2}}$$

where t is based on $n - p = 15$ df and $s = \sqrt{\text{MSE}} = \sqrt{28.367} \approx 5.326$. Thus, we have

$$\frac{338}{6} \pm 1.753 \left[\frac{5.326}{\sqrt{6}} \right] = 56.33 \pm 3.81, \text{ or } (52.52, 60.14)$$

We estimate, with 90% confidence, that the mean durability of all Brand 2 flashlight batteries lies within the interval from 52.52 to 60.14 hours.

c. The general form of a 95% confidence interval for $(\mu_2 - \mu_3)$ is:

$$(\bar{x}_2 - \bar{x}_3) \pm t_{.025}\, s \sqrt{\frac{1}{n_2} + \frac{1}{n_3}}$$

where $t_{.025} = 2.131$ is based on $n - p = 15$ df. Substitution yields:

$$\left[\frac{338}{6} - \frac{247}{6} \right] \pm 2.131(5.326)\sqrt{\frac{1}{6} + \frac{1}{6}} = 15.17 \pm 6.55,$$
$$\text{or } (8.62, 21.72)$$

We are 95% confident that the mean durability of Brand 2 batteries exceeds the mean durability of Brand 3 batteries by between 8.62 hours and 21.72 hours.

d. The hypothesis test and confidence interval procedures require the following assumptions:

(i) The durabilities of the flashlight batteries have approximate normal distributions for Brands 1, 2, and 3.
(ii) The variance of the durability distribution is the same for the three brands of batteries.

EXAMPLE 10.8 Refer to Example 10.7. Use the Bonferroni technique to compare the pairs of means corresponding to the three brands. Use $\alpha = .05$ as the overall significance level for the comparisons.

Solution The general form of the Bonferroni confidence intervals is

$$\bar{x}_i - \bar{x}_j \pm t_{\alpha/(2c)} s \sqrt{\frac{1}{n_i} + \frac{1}{n_j}}$$

where there are $c = p(p-1)/2 = 3(3-2)/2 = 3$ pairs of treatment means to be compared, $t_{\alpha/(2c)} = t_{.05/6} = t_{.0083} \approx t_{.005} = 2.947$ with $(n - p) = 18 - 3 = 15$ degrees of freedom, and $s = \sqrt{MSE} \approx 5.326$. Substituting the values of the sample means and respective sample sizes into the general formula yields:

(i) $\bar{x}_1 - \bar{x}_2 \pm (2.947)(5.326)\sqrt{\frac{1}{n_1} + \frac{1}{n_2}}$

$\frac{298}{6} - \frac{338}{6} \pm 15.6957\sqrt{\frac{1}{6} + \frac{1}{6}}$

-6.6667 ± 9.0619 or $(-15.73, 2.40)$

The mean durability for brand 1 does not differ significantly from the mean durability for brand 2, since the interval contains 0.

(ii) $\bar{x}_1 - \bar{x}_3 \pm (2.947)(5.326)\sqrt{\frac{1}{n_1} + \frac{1}{n_3}}$

$\frac{298}{6} - \frac{247}{6} \pm 15.6957\sqrt{\frac{1}{6} + \frac{1}{6}}$

8.5 ± 9.0619 or $(-.56, 17.56)$

The mean durability for brand 1 is not significantly different from that for brand 3.

(iii) $\quad \bar{x}_2 - \bar{x}_3 \pm (2.947)(5.326)\sqrt{\dfrac{1}{n_2} + \dfrac{1}{n_3}}$

$\dfrac{338}{6} - \dfrac{247}{6} \pm 15.6957\sqrt{\dfrac{1}{6} + \dfrac{1}{6}}$

$15.1667 \pm 9.0619 \text{ or } (6.10, 24.23)$

Since both endpoints of the interval are positive, the implication is that the mean durability for brand 2 is significantly greater than the mean durability for brand 3.

EXERCISES

10.1 An appliance store wished to investigate the effect of different types of advertising on the sales of its most popular microwave oven. Each of three methods (1: newspaper advertisement with the regular oven price; 2: newspaper advertisement listing a "special" price for the oven; and 3: no newspaper advertising) was used for a period of two months, with one week intervals between the different methods. The data below give the sales (number of microwave ovens sold) for randomly selected days from each period.

Method of Advertising		
1	2	3
29	30	18
26	72	17
18	47	20
35	48	18
42	28	26
19	36	14
		16
		18

a. What type of experimental design is represented here?

b. Test to see whether there are differences in the mean number of ovens sold daily for each method of advertising. Use $\alpha = .05$.

c. Estimate the mean number of ovens sold daily when no newspaper advertising is used. Use a 90% confidence interval.

d. Estimate the difference in the mean number of ovens sold daily between Methods 1 and 2 of advertising. Use a 95% confidence interval.

e. What assumptions are required for the validity of the procedures used in parts b - d?

10.2 A company is considering the implementation of a new training program for its salesmen. They have recently conducted an experiment in which five new salesmen were randomly assigned to receive the new training method and five were randomly assigned to receive the old method. The commissions for each salesman during the week following completion of the program are shown below:

Old Training Program	New Training Program
$613	$713
587	572
493	602
427	494
393	457

a. Is there evidence (at $\alpha = .025$) of a difference in the mean commission for the two training programs? Use the methods of Chapter 10 to perform an independent samples t test.

b. Now use the same data to conduct an analysis of variance F test (at $\alpha = .05$) of the null hypothesis that there is no difference in the mean commissions for the two programs.

c. Square the values of the test statistic and critical value from part a and compare them to the respective values from part b. (Note that for $p = 2$, the ANOVA F test is equivalent to the independent samples t test.)

10.3 Refer to Exercise 10.1. Use the Bonferroni technique to compare all pairs of means corresponding to the three methods of advertising. Use $\alpha = .10$ as the overall significance level for the comparisons.

10.3 The Randomized Block Design

EXAMPLE 10.9 The following experiment was designed to investigate the effect of information received by a worker regarding the quality of work being performed. Five employees at an automobile assembly plant were asked to perform a manual dexterity test under each of three experimental conditions, which were presented in a random order (1: worker receives positive feedback and encouragement from observer; 2: worker receives no information from observer; and 3: worker receives negative feedback and criticism from observer). The experimenter recorded the number of items correctly assembled during the allotted time. What type of experimental design is this?

Solution This is a randomized block design, in which the workers represent $b = 5$ blocks of relatively homogeneous experimental units. There are $p = 3$ treatments (experimental conditions), each of which is randomly assigned once to each block.

EXAMPLE 10.10 Refer to Example 10.9.

    a. Set up the test to compare the mean number of items assembled for the three experimental conditions. (Data will be provided in the following example.)

    b. Set up the test to determine if blocking is important in this experiment, i.e., if the mean numbers of items assembled differ for the five workers.

Solution

    a. It is desired to test the null hypothesis that the mean number of items correctly assembled is the same for the three experimental conditions, against the alternative that at least two of the means are different:

$$H_0: \mu_1 = \mu_2 = \mu_3$$
$$H_a: \text{At least two of the means are different}$$

where μ_1, μ_2, and μ_3 are the population mean numbers of items correctly assembled under experimental conditions 1, 2, and 3, respectively.

The test is based on the following assumptions:

    (i) The probability distributions of the number of items assembled are approximately normal for all worker-condition combinations

    (ii) The variances of all the probability distributions are equal

The test statistic is $F = \dfrac{\text{MST}}{\text{MSE}}$

and the null hypothesis will be rejected (at significance level α) if $F > F_\alpha$, where the distribution of F is based on $p - 1 = 3 - 1 = 2$ numerator degrees of freedom and $(n - b - p + 1) = (n - 5 - 3 + 1) = (n - 7)$ denominator degrees of freedom.

    b. To compare the block means, we perform a test of

        H_0: The population mean numbers of items assembled for the five workers are equal (i.e., the five block means are equal)

against

        H_a: The mean numbers of items assembled are different for at least two of the workers (i.e., at least two of the block means differ)

The test requires the same assumptions as the test in part a, and is based on the test statistic

$$F = \dfrac{\text{MSB}}{\text{MSE}}$$

The null hypothesis is rejected for all values of F such that $F > F_\alpha$, where F has $(b - 1) = 5 - 1 = 4$ numerator degrees of freedom and $(n - p - b + 1) = (n - 3 - 5 + 1) = (n - 7)$ denominator degrees of freedom.

Analysis of Variance: Comparing More Than Two Means

EXAMPLE 10.11 Refer to Examples 10.9 and 10.10. The data for the experiment are shown below:

Worker	Experimental Condition			Totals
	1	2	3	
1	32	30	25	87
2	25	26	21	72
3	19	17	14	50
4	15	12	10	37
5	12	10	8	30
Totals	103	95	78	276

a. Perform the test to compare the mean number of items assembled for the three experimental conditions. Use $\alpha = .05$.

b. Perform the test to determine if blocking is important in this experiment. Use $\alpha = .05$.

Solution

a. The following calculations are required:

$$\sum x^2 = 32^2 + 25^2 + 19^2 + \cdots + 14^2 + 10^2 + 8^2 = 5{,}914$$

$$CM = \frac{(\sum x)^2}{n} = \frac{(276)^2}{15} = 5{,}078.4$$

$$SS(\text{Total}) = \sum x^2 - CM = 5{,}914 - 5{,}078.4 = 835.6$$

$$SST = \frac{T_1^2}{b} + \frac{T_2^2}{b} + \frac{T_3^2}{b} - CM = \frac{(103)^2}{5} + \frac{(95)^2}{5} + \frac{(78)^2}{5} - 5{,}078.4$$
$$= 65.2$$

$$SSB = \frac{B_1^2}{p} + \frac{B_2^2}{p} + \frac{B_3^2}{p} + \frac{B_4^2}{p} + \frac{B_5^2}{p} - CM$$
$$= \frac{(87)^2}{3} + \frac{(72)^2}{3} + \frac{(50)^2}{3} + \frac{(37)^2}{3} + \frac{(30)^2}{3} - 5{,}078.4 = 762.27$$

$$SSE = SS(\text{Total}) - SST - SSB = 835.6 - 65.2 - 762.27 = 8.13$$

$$MST = \frac{SST}{p-1} = \frac{65.2}{2} = 32.6$$

$$MSB = \frac{SSB}{b-1} = \frac{762.27}{4} = 190.57$$

$$MSE = \frac{SSE}{n-p-b+1} = \frac{8.13}{15-3-5+1} = \frac{8.13}{8} = 1.02$$

Then the computed value of the test statistic is

$$F = \frac{MST}{MSE} = \frac{32.60}{1.02} = 31.96$$

The critical value of F is based on $p - 1 = 2$ numerator df and $n - p - b + 1 = 8$ denominator df. Thus, for $\alpha = .05$, the rejection region consists of values of F such that $F > F_{.05} = 4.46$.

Since the value of the test statistic lies within the rejection region, we conclude that the population mean numbers of items correctly assembled differ for at least two of the experimental conditions.

b. The test statistic is

$$F = \frac{\text{MSB}}{\text{MSE}} = \frac{190.57}{1.02} = 186.83$$

Since the calculated $F = 186.83$ greatly exceeds the critical value of $F_{.05} = 3.84$ ($b - 1 = 4$ numerator df and $n - p - b + 1 = 8$ denominator df), we have strong evidence that the mean numbers of items assembled differ among the five workers. Thus, the decision to use a randomized block design was wise.

EXAMPLE 10.12 Refer to Examples 10.9 - 10.11. Summarize the results of the analysis of variance in an ANOVA table.

Solution The ANOVA summary table for this randomized block design is as follows:

Source	df	SS	MS	F
Conditions (Treatments)	2	65.20	32.60	31.96
Workers (Blocks)	4	762.27	190.57	186.83
Error	8	8.13	1.02	
Totals	14	835.60		

EXAMPLE 10.13 Refer to Example 10.11. Construct a 95% confidence interval for $(\mu_1 - \mu_3)$, the difference in mean numbers of items assembled under conditions 1 and 3.

Solution The general form of a 95% confidence interval for $(\mu_1 - \mu_3)$ is

$$(\bar{x}_1 - \bar{x}_3) \pm t_{.025} s \sqrt{\frac{1}{b} + \frac{1}{b}}$$

where $b = 5$ is the number of blocks, $s = \sqrt{\text{MSE}}$, and the distribution of t is based on $n - p - b + 1 = 8$ degrees of freedom. For this example, we have

$$\bar{x}_1 = \frac{103}{5} = 20.6$$

$$\bar{x}_3 = \frac{78}{5} = 15.6$$

$$t_{.025} = 2.306$$

and

$$s = \sqrt{\text{MSE}} = \sqrt{1.02} \approx 1.01$$

Substitution yields the desired confidence interval:

$$(20.6 - 15.6) \pm 2.306(1.01)\sqrt{\frac{1}{5} + \frac{1}{5}} = 5.0 \pm 1.47, \text{ or } (3.53, 6.47)$$

We estimate, with 95% confidence, that the population mean number of items correctly assembled under condition 1 is between 3.53 and 6.47 higher than the mean number assembled correctly under condition 3.

EXAMPLE 10.14 Three of the currently most popular television shows produced the following ratings (percentage of the television audience tuned in to the show) over a period of four weeks:

	Show			
Week	A	B	C	Totals
1	33.7	27.4	22.8	83.9
2	37.1	31.2	19.7	88.0
3	34.1	31.4	24.8	90.3
4	29.4	27.2	27.9	84.5
Totals	134.3	117.2	95.2	346.7

a. Is there evidence (at $\alpha = .01$) that the mean ratings differ for the three shows?

b. Is there evidence (at $\alpha = .01$) that the use of weeks as blocks is justified in this experiment?

c. Construct a 95% confidence interval for the difference in mean ratings between Shows B and C.

d. State the assumptions necessary for the validity of the procedures used in parts a - c.

Solution

a. The hypothesis test has the following elements:

$H_0: \mu_1 = \mu_2 = \mu_3$
$H_a:$ At least two of the means differ

where μ_1, μ_2, and μ_3 are the population mean ratings for Shows A, B, and C, respectively.

At $\alpha = .01$, the null hypothesis will be rejected for all values of the test statistic F such that $F > F_{.01} = 10.92$, where F is based on $p - 1 = 3 - 1 = 2$ numerator df and $n - p - b + 1 = 12 - 3 - 4 + 1 = 6$ denominator df.

The following computations are required:

$$\sum x^2 = (33.7)^2 + (37.1)^2 + \cdots + (24.8)^2 + (27.9)^2 = 10{,}290.65$$

$$CM = \frac{(\sum x)^2}{n} = \frac{(346.7)^2}{12} = 10{,}016.74$$

$$SS(Total) = \sum x^2 - CM = 10{,}290.65 - 10{,}016.74 = 273.91$$

$$SST = \frac{T_1^2}{b} + \frac{T_2^2}{b} + \frac{T_3^2}{b} - CM$$

$$= \frac{(134.3)^2}{4} + \frac{(117.2)^2}{4} + \frac{(95.2)^2}{4} - 10{,}016.74 = 192.10$$

$$SSB = \frac{B_1^2}{p} + \frac{B_2^2}{p} + \frac{B_3^2}{p} + \frac{B_4^2}{p} - CM$$

$$= \frac{(83.9)^2}{3} + \frac{(88.0)^2}{3} + \frac{(90.3)^2}{3} + \frac{(84.5)^2}{3} - 10{,}016.74 = 9.11$$

$$SSE = SS(Total) - SST - SSB = 273.91 - 192.10 - 9.11 = 72.70$$

$$MST = \frac{SST}{p-1} = \frac{192.10}{2} = 96.05$$

$$MSB = \frac{SSB}{b-1} = \frac{9.11}{3} = 3.04$$

$$MSE = \frac{SSE}{n-p-b+1} = \frac{72.70}{12-3-4+1} = \frac{72.70}{6} = 12.12$$

Now the test statistic is

$$F = \frac{MST}{MSE} = \frac{96.05}{12.12} = 7.92$$

This value does not lie within the rejection region. There is insufficient evidence (at $\alpha = .01$) to conclude that there are differences in the mean ratings for the three shows.

b. The test statistic is

$$F = \frac{MSB}{MSE} = \frac{3.04}{12.12} = 0.25$$

and the rejection region consists of values of F such that $F > F_{.01} = 9.78$, based on $(b-1) = 3$ numerator df and $(n-p-b+1) = 6$ denominator df. The sample does not provide strong enough evidence to indicate that blocking is important in this experiment.

c. The general form of a 95% confidence interval for $(\mu_2 - \mu_3)$ is

$$(\bar{x}_2 - \bar{x}_3) \pm t_{.025} s \sqrt{\frac{1}{b} + \frac{1}{b}}$$

where

$$\bar{x}_2 = \frac{117.2}{4} = 29.3$$

$$\bar{x}_3 = \frac{95.2}{4} = 23.8$$

$$t_{.025} = 2.447 \text{ (6 df)}$$

and $\quad s = \sqrt{MSE} = \sqrt{12.12} = 3.48$

Substitution yields:

$$(29.3 - 23.8) \pm 2.447(3.48)\sqrt{\frac{1}{4} + \frac{1}{4}} = 5.5 \pm 6.02, \text{ or } (-.52, 11.52)$$

This interval is very wide and includes the value zero. Thus, we cannot conclude that the population mean ratings for Shows B and C are different. This is consistent with the F test of part a, in which we failed to reject the null hypothesis of the equality of the population means.

d. The hypothesis test and confidence interval procedures require the following assumptions:

   (i) The probability distributions of television ratings corresponding to all the show-week combinations are approximately normal.
   (ii) The variances of all the probability distributions are equal.

EXERCISES

10.4 A restaurant owner operates three restaurants within a city: one in a major shopping center (A), one near the college campus (B); and one at the beach area (C). The management has collected the following data on daily sales (in hundreds of dollars):

	Restaurant		
Day	A	B	C
Wednesday	9.5	7.4	4.9
Thursday	7.4	8.3	6.1
Friday	11.6	10.4	5.7
Saturday	17.3	6.9	13.2
Sunday	9.8	5.3	12.7

a. What type of experimental design is represented here?

b. Construct an ANOVA summary table for this experiment.

c. Is there evidence of a difference in the mean sales among the restaurants? Use $\alpha = .05$.

d. Is there evidence (at $\alpha = .05$) of a difference in the mean sales for the five days?

e. Estimate the difference in mean sales between the restaurants located at the shopping center and near the college campus. Use a 90% confidence interval.

f. State the assumptions required for the validity of the procedures used in parts b – e.

10.5 A comparison of tire prices (in dollars) for three different brands of tires (size C78-13 whitewall) produced the following data:

Type of Tire	Brand		
	A	B	C
4-ply polyester	45	48	43
Glass-belted	58	57	54
Steel-belted	71	75	68

a. Set up the ANOVA summary table.

b. Is there evidence of a difference in mean prices among the three brands? Use $\alpha = .05$.

c. Estimate the difference in the mean prices between Brands A and C with a 95% confidence interval.

10.4 Factorial Experiments

EXAMPLE 10.15 A graphic design company employs four typesetters, each of whom is trained to work at each of three machines. In an effort to investigate the possibility of interaction between typesetter and machine, each typesetter was assigned to each machine for two 1-week periods. The average number of errors per page produced during the trials were recorded. What type of experimental design does this represent?

Solution The design is a two-factor factorial experiment where factor A (typesetter) is at $a = 4$ levels and factor B (machine) is at $b = 3$ levels. The complete 4×3 factorial experiment is replicated $r = 2$ times.

EXAMPLE 10.16 Refer to Example 10.15. The data for the experiment are shown in the table.

Typesetter	Machine					
	1		2		3	
1	5	5	15	20	10	10
2	10	16	6	6	14	18
3	24	16	12	6	12	8
4	8	6	6	9	4	4

a. Perform an analysis of variance for the data and construct an ANOVA table.

b. Do the data present sufficient evidence to indicate an interaction between typesetter and type of machine? Use $\alpha = .05$.

Solution   a. The following preliminary calculations are required:

$$CM = \frac{\left(\sum x_i\right)^2}{n} = \frac{(5 + 5 + \cdots + 4 + 4)}{24} = \frac{(250)^2}{24} = 2{,}604.16667$$

$$SS(\text{Total}) = \sum x_i^2 - CM = 5^2 + 5^2 + \cdots + 4^2 + 4^2 - CM$$
$$= 3{,}292 - 2{,}604.16667 = 687.83333$$

$$SS(A) = \frac{\sum A_i^2}{br} - CM = \frac{65^2 + 70^2 + 78^2 + 37^2}{3(2)} - CM$$
$$= \frac{16{,}578}{6} - 2{,}604.16667 = 158.83333$$

$$SS(B) = \frac{\sum B_i^2}{ar} - CM = \frac{90^2 + 80^2 + 80^2}{4(2)} - CM$$
$$= \frac{20{,}900}{8} - 2{,}604.16667 = 8.33333$$

$$SS(AB) = \frac{\sum\sum AB_{ij}^2}{r} - SS(A) - SS(B) - CM$$
$$= \frac{10^2 + 35^2 + 20^2 + \cdots + 14^2 + 15^2 + 8^2}{2} - SS(A) - SS(B) - CM$$
$$= \frac{6{,}378}{2} - 158.83333 - 8.33333 - 2{,}604.16667$$
$$= 417.66667$$

$$SSE = SS(\text{Total}) - SS(A) - SS(B) - SS(AB)$$
$$= 687.83333 - 158.83333 - 8.33333 - 417.66667 = 103.00000$$

$$MS(A) = \frac{SS(A)}{a - 1} = \frac{158.83333}{3} = 52.94444$$

$$MS(B) = \frac{SS(B)}{b - 1} = \frac{8.33333}{2} = 4.16667$$

$$MS(AB) = \frac{SS(AB)}{(a - 1)(b - 1)} = \frac{417.66667}{(3)(2)} = 69.61111$$

$$MSE = \frac{SSE}{ab(r - 1)} = \frac{103.00000}{(4)(3)(1)} = 8.58333$$

The results are summarized in the following ANOVA table.

Source	df	SS	MS
Typesetter (A)	3	158.83333	52.94444
Machine (B)	2	8.33333	4.16667
Typesetter-Machine Interaction (AB)	6	417.66667	69.61111
Error	12	103.00000	8.58333
Total	23	687.83333	

b. The hypotheses of interest are:

H_0: Typesetter and Machine do not interact in their effect on average number of errors per page
H_a: There is interaction between the factors Typesetter and Machine

The test statistic is the ratio of the mean squares for interaction and error:

$$F = \frac{MS(AB)}{MSE} = \frac{69.61111}{8.58333} = 8.11$$

From the ANOVA table constructed in part a, we see that the degrees of freedom for the numerator and denominator of the F statistic are 6 and 12, respectively. Thus, at significance level .05, we will reject H_0 if $F > 3.00$.

Since the computed value of the test statistic falls within the rejection region (8.11 > 3.00), we reject H_0. There is sufficient evidence to conclude that Typesetter and Machine interact. We will consider pairwise comparisons of the sample means in the next section.

EXAMPLE 10.17 Refer to Examples 10.15 and 10.16.

a. Construct a 95% confidence interval for the average number of errors per page made by typesetter 2 using machine 3.

b. Construct a 90% confidence interval for the difference between the mean numbers of errors per page for typesetters 2 and 3 working at machine 3.

Solution

a. A 95% confidence interval for the mean $E(x)$ associated with typesetter 2 using machine 3 is

$$\bar{x}_{2,3} \pm t_{.025}\left(\frac{s}{\sqrt{r}}\right)$$

where $\bar{x}_{2,3}$ is the mean of the $r = 2$ values given for the factor level combination of typesetter 2 and machine 3, $s = \sqrt{MSE} = \sqrt{8.58333} = 2.930$, and $t_{.025} = 2.179$ based on 12 degrees of freedom. Substitution of these values into the confidence interval formula yields

$$\frac{32}{2} \pm 2.179\left(\frac{2.930}{\sqrt{2}}\right)$$

$$16 \pm 4.51$$

or (11.49, 20.51)

We are 95% confident that the mean number of errors per page made by typesetter 2 working at machine 3 is between 11.49 and 20.51.

b. A 90% confidence interval for the difference between the mean numbers of errors per page for typesetters 2 and 3 working at machine 3 is given by

$$(\bar{x}_{2,3} - \bar{x}_{3,3}) \pm t_{.05} s\sqrt{2/r}$$

where $\bar{x}_{2,3}$ and $\bar{x}_{3,3}$ are the means of the values of x obtained by typesetter 2 and 3, respectively, at machine 3, $t_{.05} = 1.782$ based on 12 degrees of freedom, and $s = 2.930$ as in part a. Substitution yields the interval

$$\left(\frac{32}{2} - \frac{20}{2}\right) \pm 1.782(2.930)\sqrt{2/2}$$
$$6 \pm 5.22$$
or $(.78, 11.22)$

Since all the values in the interval are positive, we are 90% confident that, when working on machine 3, typesetter 2 makes more errors per page, on average, than does typesetter 3.

EXERCISES

10.6 A company conducted an experiment to determine the effects of three types of incentive pay plans on worker productivity for both union and non-union workers. The company used plants in adjacent towns; one was unionized and the other was not. One-third of the production workers in each plant were assigned to each incentive plan. Then six workers were randomly selected from each group and their productivity (in numbers of items produced) was measured for a 1-week period. The six productivity measures for the factor level combinations are listed in the accompanying table.

	Incentive Plan					
	A		B		C	
Union	337	328	346	373	317	341
	362	319	351	338	335	329
	305	344	355	365	310	315
Non-union	359	346	371	377	350	336
	345	396	352	401	349	351
	381	373	399	378	374	340

a. Perform an analysis of variance for the data and construct an ANOVA table.

b. Do the data present sufficient evidence to indicate an interaction between union affiliation and incentive plan? Test using $\alpha = .05$.

10.7 Refer to Exercise 10.6.

a. Construct a 90% confidence interval for the mean productivity for a unionized worker on incentive plan B.

b. Find a 90% confidence interval for the difference in mean productivity between union and non-union workers on incentive plan B.

CHAPTER ELEVEN

Nonparametric Statistics

Summary

This chapter presented several **nonparametric techniques** for comparing two or more populations. Such methods have wide applicability and are particularly useful when observations cannot be assigned specific values, but can be ranked. The techniques require fewer restrictive assumptions than their parametric counterparts, and allow for a comparison of the probability distributions, rather than specific parameters, of the populations of interest.

The sign test is a procedure for testing hypotheses about the median of a nonnormal distribution. The **Wilcoxon rank sum test** may be used to compare two populations when data arise from an independent sampling design. When a paired difference design is employed, the **Wilcoxon signed rank test** is appropriate. The **Kruskal-Wallis H test** uses data from a completely randomized design to compare p populations. The **Friedman F_r test** is appropriate for comparing p populations based on a randomized block design. **Spearman's rank correlation coefficient** provides a nonparametric measure of correlation between two variables.

11.1 Single-Population Inferences: The Sign Test

EXAMPLE 11.1 The Environmental Protection Agency (EPA) sets certain pollution guidelines for major industries. The EPA criterion for a particular company that discharges waste into a nearby river is that the median amount of pollution in water samples collected from the river may not exceed 5 parts per million (ppm). Responding to numerous complaints, the EPA takes 10 water samples from the river at the discharge point and measures the pollution level in each sample. The results (in ppm) are given below:

5.1   4.3   5.3   6.2   5.6   4.7   8.4   5.9   6.8   3.0

Do the data provide sufficient evidence to indicate that the median pollution level in water discharged at the plant exceeds 5 ppm? Test using $\alpha = .05$.

Solution We want to test

H_0: $M = 5$
H_a: $M > 5$

using the sign test. The test statistic is

S = Number of sample observations that exceed 5
   = 7

where S has a binomial distribution with parameters $n = 10$ and $p = .5$.

From Section 8.3 we know that the observed significance level (p-value) of the test is the probability that we observe a value of the test statistic S that is at least as contradictory to the null hypothesis as the computed value. For this one-sided case, the p-value is the probability that we observe a value of S greater than or equal to 7. We find the probability using the binomial table for $n = 10$ and $p = .5$ in Table II of Appendix B:

$$p\text{-value} = P(S \geq 7) = 1 - P(S \leq 6)$$
$$= 1 - .8281$$
$$= .1719$$

Since the p-value, .1719, is larger than $\alpha = .05$, we cannot reject the null hypothesis. That is, there is insufficient evidence to indicate that the median pollution level of water discharged from the plant exceeds 5.

EXERCISE

11.1 **Scram** is the term used by nuclear engineers to describe a rapid emergency shutdown of a nuclear reactor. The nuclear industry has made a concerted effort to reduce significantly the number of unplanned scrams each year. The number of unplanned scrams at each of a random sample of 20 nuclear reactor units in a recent year are given below.

```
1  8  0  3  3  9  1  2  3  5
4  3  1  2  7  10 2  6  3  0
```

Test the hypothesis that the median number of unplanned scrams at nuclear reactor plants is less than 5. Use $\alpha = .10$.

11.2 Comparing Two Populations: The Wilcoxon Rank Sum Test for Independent Samples

EXAMPLE 11.2 The amount of financial aid available to students is often an important factor in the selection of a college or university. Nine students from a state-supported university and seven students from a private university in the same state were randomly selected and the amount of financial aid received during the last academic year was recorded for each; the results are shown in the following table:

State-Supported University	Private University
$2,400	$2,900
2,500	3,500
2,200	4,000
2,100	5,600
3,000	3,800
2,600	3,200
2,400	2,800
2,700	
3,400	

a. What assumptions would be required for the valid application of the independent samples t test to compare the population mean amounts of financial aid awarded to students at the two schools? Do you think the assumptions are reasonable in this situation?

b. Use the Wilcoxon rank sum test to see if there is a difference in the probability distributions of financial aid amounts at the two schools.

Solution

a. The independent samples t test for the difference between two population means requires the assumptions of independent samples selected from normal populations with equal variances. In this case, the assumption of normality of the population distributions may not be reasonable, since a histogram shows that the sample distributions of financial aid amounts are markedly skewed to the right. It may thus be advisable to perform the nonparametric counterpart of the test.

b. The Wilcoxon rank sum test for independent samples is a test of

H_0: The probability distributions of financial aid amounts at the two schools are identical

against

H_a: The probability distributions of the financial aid amounts are different

At $\alpha = .05$, the null hypothesis will be rejected if

$$T_B \leq 41 \text{ or } T_B \geq 78$$

where T_B is the rank sum of the amounts from the smaller sample (private university), and the critical values are obtained from part a of Table XII in the appendix of the text.

To compute the value of the test statistic, T_B, we first rank all the sample observations as though they were selected from the same population:

State-Supported University		Private University	
Observation	Rank	Observation	Rank
$2,400	3.5	$2,900	9
2,500	5	3,500	13
2,200	2	4,000	15
2,100	1	5,600	16
3,000	10	3,800	14
2,600	6	3,200	11
2,400	3.5	2,800	8
2,700	7		
3,400	12		

[Note that the two observations of $2,400 would have received ranks 3 and 4; thus, each is assigned their average rank of $(3 + 4)/2 = 3.5$.]

Now, the rank sum corresponding to the smaller sample is

$$T_B = 9 + 13 + 15 + 16 + 14 + 11 + 8 = 86$$

This value of the test statistic lies in the rejection region; thus, we conclude that the distributions of financial aid amounts awarded at the two schools are different.

EXAMPLE 11.3 Use the following data to test the hypothesis that the population distributions corresponding to A and B are identical, against the alternative that observations from population A tend to be smaller than the observations from population B. Use $\alpha = .05$.

A	B
2	12
8	7
7	6
12	7
5	9
0	4
3	4
	1

Solution We will perform a Wilcoxon rank sum test of

H_0: The probability distributions corresponding to populations A and B are identical

against

H_a: The probability distribution for population A lies below (to the left of) that for population B

At $\alpha = .05$, we will reject the null hypothesis if $T_A \leq 41$, where T_A is the rank sum for the smaller sample, and the critical value for this one-sided test is obtained from part **b** of Table XII in the text.

We now pool the measurements from both samples and rank the measurements from smallest to largest:

A		B	
Observation	Rank	Observation	Rank
2	3	12	14.5
8	12	7	10
7	10	6	8
12	14.5	7	10
5	7	9	13
0	1	4	5.5
3	4	4	5.5
		1	2

Note that the tied observations are treated as follows: The two observations of 4 would have received ranks 5 and 6; thus, each is assigned the average rank of (5 + 6)/2 = 5.5. Similarly, the three observations tied at 7 would have received ranks 9, 10, and 11; thus, each observation receives the average rank of (9 + 10 + 11)/3 = 10.

The test statistic is the rank sum associated with the smaller sample:

$$T_A = 3 + 12 + 10 + 14.5 + 7 + 1 + 4 = 51.5$$

Since this value does not fall within the rejection region, there is insufficient evidence at the $\alpha = .05$ level to support the alternative hypothesis. We cannot conclude, on the basis of this sample information, that the distribution for population A lies to the left of the distribution for population B.

EXERCISE

11.2 An elementary school teacher has experimented with two different methods of teaching long division. The final examination scores for seven randomly selected students who were taught by Method A and six who were taught by Method B are recorded below:

Method A	Method B
93	92
57	90
68	68
45	57
79	68
77	80
63	

The teacher wishes to compare the final examination scores for students taught by the two methods.

a. What assumptions are necessary to perform an independent samples t test for a difference in the mean final examination scores attained by students taught by Methods A and B? Are the assumptions reasonable?

b. Use the Wilcoxon rank sum test to determine if there is a difference (at significance level $\alpha = .10$) in the probability distributions of final examination scores for students taught by the two methods.

Nonparametric Statistics

11.3 Comparing Two Populations: Wilcoxon Signed Rank Test for the Paired Difference Experiment

EXAMPLE 11.4 Many large supermarket chains now produce their own goods for sale under a house label. They advertise that, although the house brands are of the same quality as the national brands, they can be sold at lower prices because of lower production costs. A comparison of the daily sales (number of units sold) of eleven products at a local supermarket produced the data shown in the following table.

		National Brand	House Brand
1.	Catsup	303	237
2.	Corn (canned)	504	428
3.	Bread	205	127
4.	Margarine	157	136
5.	Dog food	205	49
6.	Peaches (canned)	273	302
7.	Cola	394	147
8.	Green beans (frozen)	93	248
9.	Ice cream	188	188
10.	Cheese	126	147
11.	Beer	303	29

It is desired to compare the sales of the national brand and house brand products.

a. What type of experimental design is represented here?

b. Perform the (nonparametric) Wilcoxon signed rank test of the hypothesis that the probability distributions of the sales of national and house brands are identical. The alternative hypothesis of interest is that the sales of national brands tend to exceed those of house brands. Use $\alpha = .01$.

Solution

a. This is a paired difference experiment, and the analysis will be based on the differences between the pairs of measurements.

b. The Wilcoxon signed rank test for the paired difference design provides a test of

H_0: The probability distributions of the sales for the national and house brand products are identical

against

H_a: The sales for national brands tend to exceed those for house brands

To compute the value of the test statistic, we first obtain the ranks of the absolute values of the differences between the measurements; the calculations are shown in the following table:

Product	Sales National Brand	Sales House Brand	Difference (National−House)	Absolute Value of Difference	Rank of Absolute Value
1	303	237	66	66	4
2	504	428	76	76	5
3	205	127	78	78	6
4	157	136	21	21	1.5
5	205	49	156	156	8
6	273	302	−29	29	3
7	394	147	247	247	9
8	93	248	−155	155	7
9	188	188	0	0	(Eliminated)
10	126	147	−21	21	1.5
11	303	29	274	274	10

[Observe that differences of zero are eliminated, since they do not contribute to the rank sums. In addition, ties in absolute differences receive the average of the ranks they would be assigned if they were unequal but successive measurements. Thus, the absolute differences tied at 21, which would have received ranks 1 and 2, are each assigned the average rank of $(1 + 2)/2 = 1.5$.]

For this one-sided test, the test statistic is T_-, the negative rank sum. This is because, if the alternative hypothesis is true, we expect most of the national minus house brand sale differences to be positive; thus, we would expect the **negative** rank sum T_- to be small if the alternative hypothesis is true. The critical value will be based on $n = 10$ paired observations and is obtained from Table XII in the appendix of the text. Thus, at significance level $\alpha = .01$, the null hypothesis will be rejected if $T_- \leq 5$.

We now compute the negative rank sum, the sum of the ranks of the negative differences:

$$T_- = 3 + 7 + 1.5 = 11.5$$

This value does not lie within the rejection region; there is insufficient evidence (at $\alpha = .01$) to conclude that the sales of national brands significantly exceed sales of the house brands.

Nonparametric Statistics

EXERCISE

11.3 Refer to Exercise 9.6, in which the costs of textbooks at the campus bookstore and the off-campus bookstore were compared. The data are reproduced below for convenience.

Textbook	Campus Bookstore Price	Off-Campus Bookstore Price
1	$42.95	$40.95
2	38.00	37.00
3	35.50	36.00
4	41.00	39.95
5	46.00	44.95
6	38.00	37.50
7	43.50	42.00
8	34.95	33.50
9	37.95	35.75
10	42.00	40.50

Perform an appropriate nonparametric statistical test of

H_0: The probability distributions of textbook prices are identical for the two bookstores

against

H_a: Textbook prices at the off-campus bookstore tend to be lower than the prices at the campus bookstore

Use a significance level of $\alpha = .05$.

11.4 Kruskal-Wallis H Test for a Completely Randomized Design

EXAMPLE 11.5 A major domestic airline has initiated a campaign to make flying a more pleasant experience for its passengers. Previous passenger surveys conducted by the airline have indicated that the aspect of air travel most in need of improvement is baggage handling.

Airport managers at four of the nation's airports provided the following data on time required (in minutes) to deliver baggage from the airplane to the baggage claim area after the arrival of randomly selected flights:

Los Angeles	Chicago	Atlanta	Boston
14.2	11.9	10.8	13.1
12.8	12.2	12.1	11.9
11.9	12.9	13.1	13.2
15.3	11.8	12.8	13.0
13.9	12.8	11.9	12.5
13.0	13.1		

a. What type of experimental design is represented here? What assumptions are necessary for the analysis of these data using the methods of Chapter 10?

b. Use a nonparametric procedure to test for a difference in the probability distributions of time required to deliver baggage to the baggage claim area in the four airports. Use $\alpha = .05$.

Solution

a. The data are from a completely randomized design, in which independent random samples were selected from each of the four populations of baggage delivery times to be compared. The analysis of variance F test of Chapter 10 requires the assumption that the four distributions are approximately normal, with equal variances. If we are unwilling to make these restrictive assumptions, then a nonparametric procedure may be preferred.

b. We will perform a Kruskal-Wallis test of

H_0: The probability distributions of the populations of baggage delivery times are identical for the four airports

against

H_a: At least two of the distributions differ in location

At significance level α, the null hypothesis will be rejected if the value of the test statistic, H, exceeds χ_α^2 with $(p - 1)$ degrees of freedom, where p is the number of independent samples upon which the test is based. Thus, in our example, with $\alpha = .05$ and $p - 1 = 3$, H_0 will be rejected if $H > 7.81473$. To compute the value of H, it is necessary to obtain the rank sum for each of the four samples, where the rank of each observation is computed according to its relative magnitude when all four samples are combined. The rankings are shown in the following table.

| Los Angeles | | Chicago | | Atlanta | | Boston | |
Time	Rank	Time	Rank	Time	Rank	Time	Rank
14.2	21	11.9	4.5	10.8	1	13.1	17
12.8	11	12.2	8	12.1	7	11.9	4.5
11.9	4.5	12.9	13	13.1	17	13.2	19
15.3	22	11.8	2	12.8	11	13.0	14.5
13.9	20	12.8	11	11.9	4.5	12.5	9
13.0	14.5	13.1	17				
	$R_1 = 93.0$		$R_2 = 55.5$		$R_3 = 40.5$		$R_4 = 64.0$

(Note that tied observations are handled in the usual manner, by assigning the average value of the ranks to each of the tied observations.)

The test statistic is $H = \dfrac{12}{n(n + 1)} \sum \dfrac{R_j^2}{n_j} - 3(n + 1)$

where, for our example, $n_1 = 6$, $n_2 = 6$, $n_3 = 5$, $n_4 = 5$, $p = 4$, and $n = n_1 + n_2 + n_3 + n_4 = 22$. Substitution of these values and the rank sums computed in the table yields:

$$H = \frac{12}{22(23)} \left[\frac{(93.0)^2}{6} + \frac{(55.5)^2}{6} + \frac{(40.5)^2}{5} + \frac{(64.0)^2}{5} \right] - 3(23)$$
$$= 73.57 - 69 = 4.57$$

Since the computed value of H does not exceed the critical value of 7.81473, there is insufficient evidence to support the alternative hypothesis that the probability distributions of baggage delivery times differ in location for at least two of the airports.

EXERCISE

11.4 A consumer testing agency has recorded the lifetimes (in complete months of service) before failure of the picture tubes for random samples of three name-brand television sets. The data are presented below:

Brand A	Brand B	Brand C
32	41	48
25	39	44
40	36	43
31	47	51
35	45	41
29	34	52
37		
39		

a. Discuss the experimental design employed here. What assumptions are required for a parametric test of the hypothesis of equal means for the three lifetime distributions?

b. Use a nonparametric procedure to test for a difference in the probability distributions of the picture tube lifetimes for the three brands of television sets. Use a significance level of $\alpha = .10$.

11.5 The Friedman F_r Test for a Randomized Block Design

EXAMPLE 11.6 In the belief that most people are influenced by a product's price, label, and advertising, a local bar invited college students to participate in a beer taste test. Each rater was given three glasses of beer, in a randomized order. Glasses I and II contained the same, inexpensive local brand of beer, but students were told that Beer I was very expensive and that Beer II was not. Glass III contained a very expensive imported beer, but students were told nothing about it. Tasters were asked to rate each of the three beers on a scale from 1 to 20, with higher values indicating better taste. Ratings submitted by eight randomly selected students are shown below:

Student	Ratings		
	Beer I	Beer II	Beer III
1	17	11	13
2	19	15	12
3	16	15	19
4	13	11	17
5	15	15	11
6	17	14	17
7	19	15	17
8	20	16	19

a. Discuss the experimental design employed here.

b. Use a nonparametric test procedure to determine if there is a difference in the probability distributions of student ratings for the three beers. (Use $\alpha = .10$.)

Solution

a. This represents a randomized block design with $b = 8$ blocks (students) and $p = 3$ treatments (types of beer). The p population means may be compared using the analysis of variance techniques of Chapter 10. However, these parametric methods require the assumptions that the populations of student ratings for the three beers have normal probability distributions and that their variances are all equal. In part b, we will perform the nonparametric counterpart, which requires no distributional assumptions.

b. The elements of the Friedman F_r test for this randomized block design are as follows:

H_0: The probability distributions of student ratings are identical for the three beers

H_a: At least two of the probability distributions differ in location

At significance level α, the null hypothesis is rejected if the test statistic, F_r, exceeds the critical value χ^2_α with $(p - 1)$ degrees of freedom. For our example, we have $p - 1 = 3 - 1 = 2$ degrees of freedom and $\alpha = .10$; thus, H_0 will be rejected if $F_r > 4.60517$. In order to compute the value of F_r, it is first required to rank the observations within each block (student), and then obtain the rank sums for each of the three treatments (beers). The results are shown in the following table:

	Beer I		Beer II		Beer III	
Student	Rating	Rank	Rating	Rank	Rating	Rank
1	17	3	11	1	13	2
2	19	3	15	2	12	1
3	16	2	15	1	19	3
4	13	2	11	1	17	3
5	15	2.5	15	2.5	11	1
6	17	2.5	14	1	17	2.5
7	19	3	15	1	17	2
8	20	3	16	1	19	2
		$R_1 = 21.0$		$R_2 = 10.5$		$R_3 = 16.5$

(Tied observations within blocks are assigned the average value of the ranks each tied observation would receive if they were unequal but successive measurements.)

The test statistic is computed as follows:

$$F_r = \frac{12}{bp(p+1)} \sum R_j^2 - 3b(p+1)$$

$$= \frac{12}{8(3)(4)}[(21.0)^2 + (10.5)^2 + (16.5)^2] - 3(8)(4)$$

$$= 102.9375 - 96 = 6.9375$$

Since the calculated value of $F_r = 6.9375$ exceeds the critical value of 4.60517, we conclude that the student rating distributions differ in location for at least two of the beers.

EXERCISE

11.5 The food editor of the newspaper in a large city wishes to compare the prices of meat for four local grocery stores. For six different types of meat, she recorded the price per pound at each of the four stores. The results are shown in the following table:

	Price Per Pound			
Meat	Store 1	Store 2	Store 3	Store 4
Ground beef	$2.29	$2.09	$2.15	$2.09
Sirloin steak	4.09	3.79	3.89	3.69
Chuck roast	2.69	2.59	2.59	2.79
Pork chops	2.78	2.59	2.69	2.69
Chicken	1.69	1.55	1.59	1.59
Ham	2.69	2.49	2.59	2.49

a. What type of experimental design is represented here?

b. Perform a nonparametric analysis to determine if there is a difference in location among the probability distributions of meat prices for the four stores. (Use $\alpha = .05$.)

11.6 Spearman's Rank Correlation Coefficient

EXAMPLE 11.7 Many universities offer a career placement service which assists students in finding jobs upon graduation. A recruiter from a pharmaceutical company recently interviewed 12 graduating seniors interested in a sales position with his firm. He then ranked the candidates with respect to their performance on the interview (1 is the least favorable ranking, 12 is the most favorable), for comparison to their respective grade point averages. Results are shown below:

GPA	Interview Rank	GPA	Interview Rank
3.91	10	2.97	5
3.73	8	2.91	7
3.52	12	2.70	6
3.41	9	2.53	2
3.27	11	2.35	3
3.03	4	2.21	1

Compute the value of Spearman's rank correlation coefficient for this sample data.

Solution The sample value of Spearman's rank correlation coefficient is computed as follows:

$$r_s = \frac{SS_{uv}}{\sqrt{SS_{uu}SS_{vv}}}$$

where the u's represent the ranks of the observations on the first variable (GPA, in our example) and the v's represent the ranks on the second variable (interview performance rating).

The rankings and preliminary computations are shown in the following table:

GPA	Rank on GPA u_i	Interview Rank v_i	u_i^2	v_i^2	$u_i v_i$
3.91	12	10	144	100	120
3.73	11	8	121	64	88
3.52	10	12	100	144	120
3.41	9	9	81	81	81
3.27	8	11	64	121	88
3.03	7	4	49	16	28
2.97	6	5	36	25	30
2.91	5	7	25	49	35
2.70	4	6	16	36	24
2.53	3	2	9	4	6
2.35	2	3	4	9	6
2.21	1	1	1	1	1
	$\sum u_i = 78$	$\sum v_i = 78$	$\sum u_i^2 = 650$	$\sum v_i^2 = 650$	$\sum u_i v_i = 627$

Nonparametric Statistics

(Note that the interview performances had already been ranked when we received the data.)

Now,
$$SS_{uv} = \sum u_i v_i - \frac{(\sum u_i)(\sum v_i)}{n} = 627 - \frac{(78)(78)}{12} = 120$$

$$SS_{uu} = \sum u_i^2 - \frac{(\sum u_i)^2}{n} = 650 - \frac{(78)^2}{12} = 143$$

$$SS_{vv} = \sum v_i^2 - \frac{(\sum v_i)^2}{n} = 650 - \frac{(78)^2}{12} = 143$$

Substitution yields: $r_s = \dfrac{SS_{uv}}{\sqrt{SS_{uu} SS_{vv}}} = \dfrac{120}{\sqrt{(143)(143)}} = \dfrac{120}{143} = .839$

The positive value of r_s indicates a general tendency for the two variables to increase together; i.e., the higher a candidate's GPA, the higher his or her interview performance ranking tends to be.

EXAMPLE 11.8 Refer to Example 11.7. Use the shortcut formula to compute Spearman's rank correlation coefficient, r_s, for the GPA-interview ranking data.

Solution The following shortcut formula may be used for computing r_s when there are no ties in the rankings of the observations:

$$r_s = 1 - \frac{6 \sum d_i^2}{n(n^2 - 1)}$$

where $d_i = u_i - v_i$, the difference in rankings on GPA and interview performance for the ith individual.

The necessary computations are shown in the following table:

GPA	Rank on GPA u_i	Interview Rank v_i	$d_i = u_i - v_i$	d_i^2
3.91	12	10	2	4
3.73	11	8	3	9
3.52	10	12	−2	4
3.41	9	9	0	0
3.27	8	11	−3	9
3.03	7	4	3	9
2.97	6	5	1	1
2.91	5	7	−2	4
2.70	4	6	−2	4
2.53	3	2	1	1
2.35	2	3	−1	1
2.21	1	1	0	0
			$\sum d_i^2 =$	46

Then we calculate

$$r_s = 1 - \frac{6\sum d_i^2}{n(n^2 - 1)} = 1 - \frac{6(46)}{12(12^2 - 1)} = 1 - .161 = .839$$

as obtained in Example 11.7

EXAMPLE 11.9 Refer to Examples 11.7 and 11.8. Perform a test (at significance level $\alpha = .01$) of the null hypothesis that the population Spearman rank correlation coefficient, ρ_s, between GPA and interview ranking is zero, against the alternative that the two variables are positively correlated.

Solution The elements of the test are the following:

H_0: $\rho_s = 0$
H_a: $\rho_s > 0$

For $\alpha = .01$, the rejection region (obtained by consulting Table XIV in the appendix of the text) consists of all values of r_s such that $r_s > .703$.

In Examples 11.7 and 11.8, we computed $r_s = .839$. This value is in the rejection region, and we conclude that GPA and interview ranking are positively correlated.

EXERCISES

11.6 A consumer magazine has published the following data on the quality and cost of the eight best-selling models of food processing machines (Rank 1 corresponds to the lowest quality, and Rank 8 to the highest):

Model	Quality Rank	Cost ($)
1	1	$ 89
2	2	99
3	3	135
4	4	120
5	5	179
6	6	249
7	7	199
8	8	225

a. Compute the value of Spearman's rank correlation coefficient, r_s, for these sample data.

b. Use the shortcut formula to calculate the value of r_s for the above data.

11.7 Refer to Exercise 11.6. Test the null hypothesis that the population Spearman rank correlation coefficient, ρ_s, is zero, against the alternative that the quality and cost of food processing machines are positively correlated. Use $\alpha = .05$.

Nonparametric Statistics

CHAPTER TWELVE

The Chi-Square Test and the Analysis of Contingency Tables

Summary

This chapter presented methods of analyzing data from experiments which give rise to **count** or **enumerative** data.

A χ^2 procedure may be applied to **one-dimensional** count data to test hypotheses that the **multinomial probabilities** are equal to specified values. When count data are classified in a **two-dimensional contingency table**, the χ^2 statistic allows us to test the independence of the two methods of classifying the data.

The chapter concluded with some words of caution regarding common misuses of the χ^2 procedure.

12.1 One-Dimensional Count Data: The Multinomial Distribution

EXAMPLE 12.1 Automotive manufacturers are now predicting that by 1995, 40% of all new cars purchased will be sub-compacts, 35% will be compacts, 20% will be mid-sized, and the remaining 5% will be luxury models. A random sample of $n = 200$ auto purchases in June 1993 showed that 63 were sub-compacts, 55 were compacts, 49 were mid-sized, and 33 were luxury. Does this sample evidence indicate a significant difference between the current and projected trends for the purchase of new automobiles? Use a significance level of $\alpha = .05$.

Solution We define the following notation:

p_1 = true proportion of current new automobile purchases which are subcompacts
p_2 = true proportion of current car purchases which are compacts
p_3 = true proportion of current new car purchases which are mid-sized
p_4 = true proportion of current new car purchases which are luxury models

The null hypothesis of interest is that the current purchasing trend is identical to that predicted for 1995; the alternative hypothesis is that there is a significant difference between current and projected trends. Thus, we have

H_0: $p_1 = .40, p_2 = .35, p_3 = .20, p_4 = .05$
H_a: At least one of the proportions differs from its hypothesized value

Since our multinomial experiment consists of $k = 4$ possible outcomes, the test procedure will be based on a χ^2 distribution with $k - 1 = 4 - 1 = 3$ degrees of freedom. Thus, at $\alpha = .05$, the null hypothesis will be rejected for all values of the test statistic X^2 such that

$$X^2 > \chi^2_{.05} = 7.81473$$

Now, if the null hypothesis were true (i.e., if $p_1 = .40$, $p_2 = .35$, $p_3 = .20$, and $p_4 = .05$), then we would expect to observe the following counts for each possible outcome of the experiment:

$$E(n_1) = np_{1,0} = 200(.40) = 80$$
$$E(n_2) = np_{2,0} = 200(.35) = 70$$
$$E(n_3) = np_{3,0} = 200(.20) = 40$$
$$E(n_4) = np_{4,0} = 200(.05) = 10$$

In order to measure the amount of disagreement between the sample data and the null hypothesis, we compute the test statistic:

$$
\begin{aligned}
X^2 &= \frac{[n_1 - E(n_1)]^2}{E(n_1)} + \frac{[n_2 - E(n_2)]^2}{E(n_2)} + \frac{[n_3 - E(n_3)]^2}{E(n_3)} + \frac{[n_4 - E(n_4)]^2}{E(n_4)} \\
&= \frac{(n_1 - 80)^2}{80} + \frac{(n_2 - 70)^2}{70} + \frac{(n_3 - 40)^2}{40} + \frac{(n_4 - 10)^2}{10} \\
&= \frac{(63 - 80)^2}{80} + \frac{(55 - 70)^2}{70} + \frac{(49 - 40)^2}{40} + \frac{(33 - 10)^2}{10} = 61.75
\end{aligned}
$$

The computed value of $X^2 = 61.75$ exceeds the critical value of $X^2 = 7.81473$. Thus, at significance level $\alpha = .05$, we reject the null hypothesis and conclude that the current trend in purchasing new automobiles differs significantly from that projected for 1995.

EXAMPLE 12.2 Refer to Example 12.1. Construct a 95% confidence interval for p_4, the proportion of current new car purchases which are for luxury models. Interpret the interval.

Solution Recall from Chapter 7 that the general form of a 95% confidence interval for p_4 is given by

$$\hat{p}_4 \pm z_{.025} \sqrt{\frac{\hat{p}_4 \hat{q}_4}{n}}$$

where, for this example,

$$\hat{p}_4 = \frac{33}{200} = .165$$
$$\hat{q}_4 = 1 - \hat{p}_4 = 1 - .165 = .835$$
$$n = 200$$
$$z_{.025} = 1.96$$

The Chi-Square Test and the Analysis of Contingency Tables

Substitution yields the desired interval:

$$.165 \pm 1.96\sqrt{\frac{(.165)(.835)}{200}} = .165 \pm .051, \text{ or } (.114, .216)$$

We estimate, with 95% confidence, that luxury models account for between 11.4% and 21.6% of current new car purchases. In other words, the projected proportion of .05 for 1995 is currently being exceeded.

EXAMPLE 12.3 A large food wholesaler has been experimenting with a new process to make "natural" peanut butter. The company conducted a consumer survey which asked 300 respondents to give their taste preference for one of the three types of peanut butter: standard "smooth" peanut butter; standard "crunchy" peanut butter; or new "natural" peanut butter. The results are shown below:

Consumer Preference		
Smooth	Crunchy	Natural
87	109	104

Is there sufficient evidence to indicate that a preference exists for one or more of the types of peanut butter? Use $\alpha = .01$.

Solution The relevant hypothesis test is composed of the following elements:

H_0: $p_1 = 1/3$, $p_2 = 1/3$, $p_3 = 1/3$
H_a: At least one proportion differs from its hypothesized value

where

p_1 = proportion of all consumers who prefer the smooth peanut butter (or equivalently, the probability a randomly selected consumer prefers the smooth peanut butter)
p_2 = proportion of all consumers who prefer the crunchy peanut butter
p_3 = proportion of all consumers who prefer the natural peanut butter

(Note that the null hypothesis states there is no preference for any particular type of peanut butter; i.e., a consumer is equally likely to prefer any of the three types.)

The multinomial experiment of selecting a type of peanut butter has $k = 3$ possible outcomes (smooth, crunchy, natural). Thus, the test procedure is based on a χ^2 distribution with $k - 1 = 2$ degrees of freedom. At significance level $\alpha = .01$, the null hypothesis will be rejected if

$$X^2 > \chi^2_{.01} = 9.21034$$

The expected counts which would be observed for each outcome, assuming the null hypothesis were true, are computed as follows:

$$E(n_1) = np_{1,0} = 300(1/3) = 100$$
$$E(n_2) = np_{2,0} = 300(1/3) = 100$$
$$E(n_3) = np_{3,0} = 300(1/3) = 100$$

The test statistic is then given by

$$X^2 = \sum \frac{[n_i - E(n_i)]^2}{E(n_i)} = \frac{(87-100)^2}{100} + \frac{(109-100)^2}{100} + \frac{(104-100)^2}{100}$$
$$= 2.66$$

Since this value of X^2 does not lie within the rejection region, there is insufficient evidence (at significance level .01) to conclude that a preference exists for one or more of the types of peanut butter.

EXERCISES

12.1 Before credit guidelines were tightened, it is known that 35% of the purchases at a large department store were paid for with a national credit card, 30% with the department store credit card, 15% with cash, and 20% with check. However, a recent survey, conducted to determine how shoppers at the department store prefer to pay for their purchases, produced the following results:

Method of Payment			
National Credit Card	Department Store Credit Card	Cash	Check
145	110	106	139

Do these survey results indicate (at $\alpha = .01$) that the proportions of shoppers who prefer the various modes of payment have changed significantly from previous years?

12.2 A newspaper article in one metropolitan city stated that 39% of the state's licensed drivers have never been involved in an accident, 27% have had one accident, and the remaining 34% have had more than one accident. A sample of 160 randomly selected licensed drivers from this state indicated that 50 had never had an accident, 39 had been involved in one accident, and 71 had more than one accident. Does the sample evidence provide sufficient information to refute the percentages reported by the newspaper? Use $\alpha = .01$.

12.2 Contingency Tables

EXAMPLE 12.4 A recent sample of 300 randomly selected purchases on the New York Stock Exchange were classified according to annual income of the buyer and the price of a single share of the purchased stock. The results are shown in the table:

		Price Per Share				
		Less than $10	$10–$29.99	$30–$49.99	$50 or More	Totals
Annual Income of Buyer	Less than $20,000	28	21	9	2	60
	$20,000–$50,000	29	43	45	24	141
	More than $50,000	4	29	38	28	99
	Totals	61	93	92	54	300

Test to see whether the buyer's annual income and price per share of stock are dependent. Use $\alpha = .05$.

Solution

We wish to conduct the following test for independence:

H_0: The price per share of a purchased stock is independent of the buyer's annual income
H_a: Price per share and annual income are dependent

The two-dimensional contingency table contains $r = 3$ rows and $c = 4$ columns. Thus, the χ^2 procedure will be based upon $(r-1)(c-1) = (3-1)(4-1) = 2(3) = 6$ degrees of freedom. At significance level $\alpha = .05$, the rejection region consists of those values of the test statistic X^2 such that

$$X^2 > \chi^2_{.05} = 12.5916$$

The next step is to compute the cell frequencies which we would expect to obtain if the null hypothesis were true, i.e., if the two classifications "price per share" and "annual income" were in fact independent.

Thus, for example,

$$\hat{E}(n_{11}) = \frac{r_1 c_1}{n} = \frac{(60)(61)}{300} = 12.20$$

$$\hat{E}(n_{12}) = \frac{r_1 c_2}{n} = \frac{(60)(93)}{300} = 18.60$$

$$\hat{E}(n_{13}) = \frac{r_1 c_3}{n} = \frac{(60)(92)}{300} = 18.40$$

$$\vdots$$

$$\hat{E}(n_{33}) = \frac{r_3 c_3}{n} = \frac{(99)(92)}{300} = 30.36$$

$$\hat{E}(n_{34}) = \frac{r_3 c_4}{n} = \frac{(99)(54)}{300} = 17.82$$

All of the observed and estimated expected (in parentheses) cell counts are shown in the table below:

		Price Per Share			
		Less than $10	$10–$29.99	$30–$49.99	$50 or more
Annual Income	Less than $20,000	28 (12.20)	21 (18.60)	9 (18.40)	2 (10.80)
	$20,000–$50,000	29 (28.67)	43 (43.71)	45 (43.24)	24 (25.38)
	More than	4 (20.13)	29 (30.69)	38 (30.36)	28 (17.82)

Then the value of the test statistic is calculated as follows:

$$X^2 = \sum \frac{[n_{ij} - \hat{E}(n_{ij})]^2}{\hat{E}(n_{ij})}$$

$$= \frac{(28 - 12.20)^2}{12.20} + \frac{(21 - 18.60)^2}{18.60} + \frac{(9 - 18.40)^2}{18.40}$$

$$+ \cdots + \frac{(29 - 30.69)^2}{30.69} + \frac{(38 - 30.36)^2}{30.36} + \frac{(28 - 17.82)^2}{17.82}$$

$$= 53.66$$

Since $X^2 = 53.66$ exceeds the critical value of $X^2 = 12.5916$, there is sufficient evidence (at the $\alpha = .05$ level) to conclude that the price per share of a purchased stock and the buyer's annual income are dependent.

EXAMPLE 12.5 A state legislature is considering a bill to reduce the penalties for possession and use of marijuana. The following table shows the results of a survey of 67 randomly selected voters, who are classified according to their educational level and opinion on the proposed legislation:

		Years of Education				
		0–8	9–12	13–16	More than 16	Totals
Opinion	Approve	6	8	12	6	32
	Disapprove	15	4	10	6	35
	Totals	21	12	22	12	67

Is there evidence (at $\alpha = .01$) of a relationship between the number of years of education and opinion on the proposed legislative bill?

Solution The relevant hypothesis test consists of the following elements:

H_0: Educational level and opinion on reducing the penalty for possession and use of marijuana are independent

H_a: Educational level and opinion on reducing the penalty for possession and use of marijuana are dependent

The Chi-Square Test and the Analysis of Contingency Tables

The test will be based upon a χ^2 distribution with $(r-1)(c-1) = (2-1)(4-1) = 3$ degrees of freedom, since the contingency table has $r = 2$ rows and $c = 4$ columns. Thus, at significance level $\alpha = .01$, the null hypothesis will be rejected if

$$X^2 > \chi^2_{.01} = 11.3449$$

Calculation of the estimated expected cell frequencies proceeds as follows:

$$\hat{E}(n_{11}) = \frac{r_1 c_1}{n} = \frac{(32)(21)}{67} = 10.03$$

$$\hat{E}(n_{12}) = \frac{r_1 c_2}{n} = \frac{(32)(12)}{67} = 5.73$$

$$\vdots$$

$$\hat{E}(n_{24}) = \frac{r_2 c_4}{n} = \frac{(35)(12)}{67} = 6.27$$

The observed and estimated expected cell frequencies are shown in the following table:

		Years of Education			
		0–8	9–12	13–16	More than 16
Opinion	Approve	6 (10.03)	8 (5.73)	12 (10.51)	6 (5.73)
	Disapprove	15 (10.97)	4 (6.27)	10 (11.49)	6 (6.27)

The test statistic is then computed as follows:

$$X^2 = \sum \frac{\left[n_{ij} - \hat{E}(n_{ij})\right]^2}{\hat{E}(n_{ij})}$$

$$= \frac{(6 - 10.03)^2}{10.03} + \frac{(8 - 5.73)^2}{5.73} + \frac{(12 - 10.51)^2}{10.51} + \frac{(6 - 5.73)^2}{5.73}$$

$$+ \frac{(15 - 10.97)^2}{10.97} + \frac{(4 - 6.27)^2}{6.27} + \frac{(10 - 11.49)^2}{11.49} + \frac{(6 - 6.27)^2}{6.27}$$

$$= 5.25$$

At significance level $\alpha = .01$, this value of the test statistic does not lie within the rejection region. We cannot conclude that educational level and opinion on the proposed legislation are dependent.

EXERCISES

12.3 A banking institution in a large city has conducted a survey to investigate attitudes about borrowing money. Each of 300 randomly selected savings account customers in the 20-25 year age group was asked to classify his or her own attitude toward borrowing, and also the attitude of his or her parents toward borrowing, into one of the following categories:

1. Never borrow money.
2. Borrow money only in emergency situations or to make large purchases.
3. Borrow money to solve occasional, temporary cash-flow problems.
4. Borrow money as necessary to meet regular obligations.

The responses are classified in the following table:

		Attitude of Parents			
		1	2	3	4
Attitude of Respondent	1	50	5	5	5
	2	30	40	10	5
	3	25	10	30	20
	4	15	15	10	25

Test to determine if there is a relationship (at $\alpha = .05$) between the attitudes of the respondents and the attitudes of parents toward borrowing money.

12.4 A survey of 84 randomly selected college seniors provided the following breakdown by sex and political party affiliation:

		Political Party Affiliation			
		Republican	Democrat	Independent	Other
Sex	Male	3	28	12	7
	Female	10	10	11	3

Is there sufficient evidence to conclude that political party affiliation and sex are related? Use a significance level of $\alpha = .01$.

The Chi-Square Test and the Analysis of Contingency Tables

CHAPTER THIRTEEN

Simple Linear Regression

Summary

This chapter presented the five steps required in a **regression analysis**, a procedure for fitting a prediction equation to a set of data and making inferences from the results. We restricted attention to the particular case where a dependent variable y is related to a single independent variable x. The five steps are summarized below:

1. A **probabilistic model** is hypothesized. Straight-line models are of the form $y = \beta_0 + \beta_1 x + \epsilon$.

2. The unknown parameters in the **deterministic component**, $\beta_0 + \beta_1 x$, are estimated using the **method of least squares**. The sum of squared errors for the resulting least squares model is smaller than that for any other straight-line model.

3. The probability distribution of ϵ, the **random error component**, is specified.

4. Inferences about the slope β_1, and the calculation of the **coefficient of correlation** r and the **coefficient of determination** r^2 are performed to assess the usefulness of the model.

5. If judged to be satisfactory, the model may be used to estimate $E(y)$, the mean y value for a given x value, or to predict an individual y value for a specific value of x.

13.1 Probabilistic Models
13.2 Fitting the Model: Least Squares Approach

EXAMPLE 13.1 a. Plot the graph of the (deterministic) straight line $y = 1 + .5x$.

b. Give the slope and y-intercept of the line defined in part a.

Solution a.

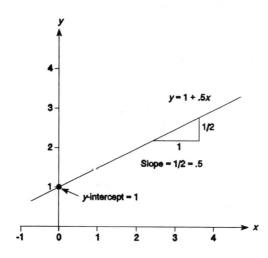

b. For a straight line of the form $y = \beta_0 + \beta_1 x$, the y-intercept is β_0 and the slope is β_1. In our example,

$$y\text{-intercept} = 1$$
$$\text{and slope} = .5$$

Note from the figure that the y-intercept (1) is the point at which the line crosses the y-axis. Also, the slope (.5) is the amount of increase in y for a unit increase in x.

EXAMPLE 13.2 The financial director of a private college was interested in the relationship between the dependent variable,

y = amount of money donated to the school per year by an alumnus

and the independent variable,

x = number of years the alumnus has been out of school

Simple Linear Regression

A random sample of 15 alumni donors yielded the following information:

Number of Years Out of School, x	Amount Donated, y ($)
4	25
8	20
30	50
24	150
18	100
10	10
21	75
7	30
15	25
28	60
2	10
17	25
12	30
21	50
14	10

a. Plot a scattergram of the data.

b. Use the method of least squares to fit a straight line to the $n = 15$ data points. Graph the least squares line on the scattergram.

Solution

a.

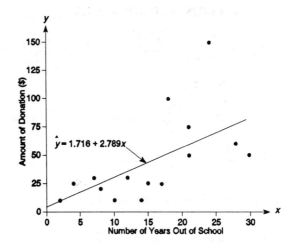

Note the scattergram suggests a general tendency for y to increase as x increases.

b. We will set up a table to assist in performing the required calculations:

x	y	x^2	xy
4	25	16	100
8	20	64	160
30	50	900	1,500
24	150	576	3,600
18	100	324	1,800
10	10	100	100
21	75	441	1,575
7	30	49	210
15	25	225	375
28	60	784	1,680
2	10	4	20
17	25	289	425
12	30	144	360
21	50	441	1,050
14	10	196	140
$\sum x = 231$	$\sum y = 670$	$\sum x^2 = 4,553$	$\sum xy = 13,095$

We now calculate:

$$SS_{xy} = \sum xy - \frac{(\sum x)(\sum y)}{n} = 13,095 - \frac{(231)(670)}{15}$$

$$= 13,095 - 10,318 = 2,777$$

$$SS_{xx} = \sum x^2 - \frac{(\sum x)^2}{n} = 4,553 - \frac{(231)^2}{15} = 4,553 - 3,557.4$$

$$= 995.6$$

$$\bar{y} = \frac{\sum y}{n} = \frac{670}{15} = 44.667$$

$$\bar{x} = \frac{\sum x}{n} = \frac{231}{15} = 15.4$$

Now the slope of the least squares line is

$$\hat{\beta}_1 = \frac{SS_{xy}}{SS_{xx}} = \frac{2,777}{995.6} = 2.789$$

and the y-intercept is

$$\hat{\beta}_0 = \bar{y} - \hat{\beta}_1 \bar{x} = 44.667 - 2.789(15.4) = 1.716$$

Thus, the least squares line is

$$\hat{y} = \hat{\beta}_0 + \hat{\beta}_1 x = 1.716 + 1.789x$$

This line is graphed on the scattergram in part a.

Simple Linear Regression

EXAMPLE 13.3 Refer to Example 13.2. Compute SSE, the sum of squared errors for the least squares model.

Solution The following table presents the calculations required for SSE:

x	y	$\hat{y} = 1.716 + 2.789x$	$(y - \hat{y})$	$(y - \hat{y})^2$
4	25	12.872	12.128	147.09
8	20	24.028	−4.028	16.22
30	50	85.386	−35.386	1,252.17
24	150	68.652	81.348	6,617.50
18	100	51.918	48.082	2,311.88
10	10	29.606	−19.606	384.40
21	75	60.285	14.715	216.53
7	30	21.239	8.761	76.76
15	25	43.551	−18.551	344.14
28	60	79.808	−19.808	392.36
2	10	7.294	2.706	7.32
17	25	49.129	−24.129	582.21
12	30	35.184	−5.184	26.87
21	50	60.285	−10.285	105.78
14	10	40.762	−30.762	946.30
				SSE = 13,427.53

This value of SSE = 13,427.53 is smaller than the SSE for any other straight-line model which could be fit to the data.

EXAMPLE 13.4 Refer to Example 13.2. According to your least squares line, approximately how much money would an alumnus who has been out of school for 20 years be expected to donate?

Solution The least squares prediction equation obtained in Example 13.2 is

$$\hat{y} = 1.716 + 2.789x$$

For an alumnus who has been out of school for 20 years (i.e., $x = 20$), the predicted amount donated (in dollars) per year is

$$\hat{y} = 1.716 + 2.789(20) = 57.496$$

We would expect the alumnus to donate approximately $57.50 per year to the school. (A measure of the reliability of our prediction will be developed in a subsequent section.)

EXERCISES

13.1 a. Plot the graph of the straight line

$$y = 1.5 - 2x$$

b. Give the slope and y-intercept of the line defined in part a. Interpret these values.

13.2 A study was conducted to investigate the relationship between a student's score on the Graduate Management Aptitude Test (GMAT) and his or her grade point average (GPA). It is hoped that a useful equation will be developed for predicting a student's GMAT score from his or her GPA. Records for ten students yielded the following data.

GMAT Score y	GPA x
710	3.8
690	3.6
650	3.7
630	3.4
490	2.7
680	3.1
500	3.0
550	2.9
660	3.5
430	2.4

a. Plot a scattergram of the data.

b. Use the method of least squares to fit a straight line to the data points. Graph the least squares line on the scattergram.

c. Compute the value of SSE for the least squares model.

d. According to your least squares prediction equation, approximately what GMAT score would you expect for a student whose GPA is 3.3?

13.3 Model Assumptions
13.4 An Estimator of σ^2

EXAMPLE 13.5 Refer to Example 13.3. Calculate s^2, the estimator of σ^2, the variance of the random error component ϵ of the probabilistic model

$$y = \beta_0 + \beta_1 x + \epsilon$$

Simple Linear Regression

Solution   In the alumnus donation example, we previously calculated SSE = 13,427.53 for the least squares line $\hat{y} = 1.716 + 2.789x$. Since there were $n = 15$ data points, there are $(n - 2) = 13$ degrees of freedom available for estimating σ^2. Thus,

$$s^2 = \frac{SSE}{n-2} = \frac{13{,}427.53}{13} = 1{,}032.8869$$

is our estimate of the variance of ϵ. The estimated standard deviation of ϵ,

$s = \sqrt{s^2}$, will later be used to evaluate the error of prediction when using the least squares line to predict a value of y for a specified value of x.

(Note that in the calculation of SSE, it may often be easier to use the computational formula

$$SSE = SS_{yy} - \hat{\beta}_1 SS_{xy}$$

since the quantities $\hat{\beta}_1$ and SS_{xy} were computed during the fitting of the least squares line.)

EXAMPLE 13.6   A major appliance store has obtained the following data on daily high temperature and number of air-conditioning units sold for eight randomly selected business days during the summer:

Daily High Temperature x, °F	Number of Units Sold y
83	4
88	5
73	1
76	0
92	5
79	3
81	2
77	2

a. Fit a least squares line to the data.

b. Plot the data and graph the least squares line.

c. Compute SSE.

d. Calculate s^2.

Solution a. The following table shows the required calculations:

x	y	x^2	xy
83	4	6,889	332
88	5	7,744	440
73	1	5,329	73
76	0	5,776	0
92	5	8,464	460
79	3	6,241	237
81	2	6,561	162
77	2	5,929	154
$\sum x = 649$	$\sum y = 22$	$\sum x^2 = 52{,}933$	$\sum xy = 1{,}858$

Thus,

$$SS_{xy} = \sum xy - \frac{(\sum x)(\sum y)}{n} = 1{,}858 - \frac{(649)(22)}{8} = 73.25$$

$$SS_{xx} = \sum x^2 - \frac{(\sum x)^2}{n} = 52{,}933 - \frac{(649)^2}{8} = 282.875$$

$$\bar{y} = \frac{\sum y}{n} = \frac{22}{8} = 2.75$$

$$\bar{x} = \frac{\sum x}{n} = \frac{649}{8} = 81.125$$

Then the slope of the least squares line is

$$\hat{\beta}_1 = \frac{SS_{xy}}{SS_{xx}} = \frac{73.25}{282.875} = .258948$$

and the y-intercept is

$$\hat{\beta}_0 = \bar{y} - \hat{\beta}_1 \bar{x} = 2.75 - .258948(81.125) = -18.257157$$

Thus, the least squares line is

$$\hat{y} = \hat{\beta}_0 + \hat{\beta}_1 x \text{ or } \hat{y} = -18.26 + .26x$$

Simple Linear Regression

b. A graph of the data points and the least squares line follows:

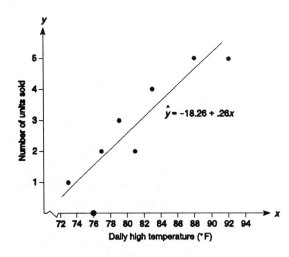

c. We will use the computational formula for SSE:

$$SSE = SS_{yy} - \hat{\beta}_1 SS_{xy}$$

where

$$SS_{yy} = \sum y^2 - \frac{(\sum y)^2}{n}$$
$$= (4^2 + 5^2 + 1^2 + 0^2 + 5^2 + 3^2 + 2^2 + 2^2) - \frac{(22)^2}{8}$$
$$= 84 - 60.5 = 23.5$$

Hence,

$$SSE = 23.5 - (.258948)(73.25) = 4.532059$$

(Note that we have retained six significant digits in the value for $\hat{\beta}_1$ to avoid substantial rounding errors in the calculation of SSE.)

d. Our estimate of the variance of the random error component ϵ is:

$$s^2 = \frac{SSE}{n-2} = \frac{4.532059}{8-2} = .755343$$

EXERCISES

13.3 Refer to Exercise 13.2. Obtain an estimate of σ^2, the variance of the random error component in the model.

13.4 Refer to Exercise 13.2. Re-compute SSE, using the computational formula:

$$SSE = SS_{yy} - \hat{\beta}_1 SS_{xy}$$

13.5 Assessing the Usefulness of the Model: Making Inferences About the Slope β_1

EXAMPLE 13.7 Refer to Examples 13.2-13.5.

a. State the assumptions about ϵ in the model

$$y = \beta_0 + \beta_1 x + \epsilon$$

where y = amount of money donated to the school per year by an alumnus and x = number of years the alumnus has been out of school.

b. Test the null hypothesis that x contributes no information (at $\alpha = .05$) for the prediction of y against the alternative that amount donated, y, tends to increase as the number of years out of school, x, increases.

Solution

a. In order to make inferences about the usefulness of the model, the following four assumptions are made about the probability distribution of ϵ:

(i) The probability distribution of ϵ is normal.
(ii) The expected value or mean of the probability distribution is zero.
(iii) The variance of the probability distribution is equal to a constant, σ^2, for all values of x.
(iv) The errors associated with different observations of y are independent.

b. The hypothesis test has the following elements:

H_0: $\beta_1 = 0$
H_a: $\beta_1 > 0$

where β_1 is the true slope of the straight line relating y and x. At $\alpha = .05$, we will reject H_0 if

$$t > t_{.05} = 1.771$$

where t is based on $n - 2 = 15 - 2 = 13$ degrees of freedom.

The value of the test statistic is:

$$t = \frac{\hat{\beta}_1 - 0}{s_{\hat{\beta}_1}} = \frac{\hat{\beta}_1 - 0}{s/\sqrt{SS_{xx}}}$$

We have previously obtained

$$\hat{\beta}_1 = 2.789$$

$$s = \sqrt{s^2} = \sqrt{1032.8869} \approx 32.14$$

and

$$SS_{xx} = 995.6$$

Simple Linear Regression

Substitution yields

$$t = \frac{2.789 - 0}{32.14/\sqrt{995.6}} = 2.74$$

The calculated t value falls within the rejection region; we thus reject H_0 and conclude that β_1 is significantly greater than zero. The sample evidence suggests that y tends to increase as x increases, and that x contributes useful information for the prediction of y.

EXAMPLE 13.8 Refer to Example 13.7. Construct a 90% confidence interval for β_1.

Solution The general form for a 90% confidence interval for β_1 is

$$\hat{\beta}_1 \pm t_{.05} s_{\hat{\beta}_1}$$

where $t_{.05} = 1.771$ is based on $n - 2 = 13$ degrees of freedom and

$$s_{\hat{\beta}_1} = s/\sqrt{SS_{xx}} \approx 32.14/\sqrt{995.6} = 1.02$$

The required confidence interval is then

$$2.789 \pm 1.771(1.02) = 2.789 \pm 1.806, \text{ or } (.983, 4.595)$$

We are 90% confident that the slope parameter β_1 is contained in this interval. That is, we estimate (with 90% confidence) that the mean increase in the amount donated per year by an alumnus for each unit increase in number of years out of school is between $.98 and $4.60.

EXAMPLE 13.9 For five popular American-made cars, the following information on engine size and mileage ratings was recorded:

Engine Size, x (cubic inches)	Mileage Rating, y (miles per gallon of gasoline)
144	28
232	21
306	23
388	17
414	15

For these data, $SS_{xx} = 49,684.8$, $SS_{yy} = 104.8$, $SS_{xy} = -2,119.2$, $\sum x = 1,484$, and $\sum y = 104$.

Test the null hypothesis that x contributes no information for the prediction of y against the alternative that these variables are linearly related with a slope significantly different from zero. Use a significance level of $\alpha = .01$.

Solution

The hypothesized probabilistic model is

$$y = \beta_0 + \beta_1 x + \epsilon$$

where y = mileage rating, x = engine size, and the errors ϵ are assumed to be independent and normally distributed with mean 0 and constant variance σ^2.

The least squares line is given by

$$\hat{y} = \hat{\beta}_0 + \hat{\beta}_1 x$$

where

$$\hat{\beta}_1 = \frac{SS_{xy}}{SS_{xx}} = \frac{-2{,}119.2}{49{,}684.8} = -.043$$

and

$$\hat{\beta}_0 = \bar{y} - \hat{\beta}_1 \bar{x} = \frac{104}{5} - (-.043)\left(\frac{1{,}484}{5}\right)$$

$$= 20.80 + 12.76 = 33.56$$

The hypotheses of interest are:

$H_0: \beta_1 = 0$
$H_a: \beta_1 \neq 0$

At $\alpha = .01$, we will reject H_0 if

$$t < -t_{.005} \text{ or } t > t_{.005}$$

i.e., if $t < -5.841$ or $t > 5.841$, where t is based on $n - 2 = 5 - 2 = 3$ degrees of freedom.

The form of the test statistic is

$$t = \frac{\hat{\beta}_1 - 0}{s_{\hat{\beta}_1}} = \frac{\hat{\beta}_1}{s/\sqrt{SS_{xx}}}$$

where

$$s = \sqrt{\frac{SSE}{n-2}} = \sqrt{\frac{SS_{yy} - \hat{\beta}_1 SS_{xy}}{n-2}} = \sqrt{\frac{104.8 - (-.043)(-2{,}119.2)}{3}}$$

$$= \sqrt{4.5581} = 2.13$$

The computed value of t is then

$$t = \frac{-.043}{2.13/\sqrt{49{,}684.8}} = -4.50$$

At $\alpha = .01$, there is insufficient evidence to conclude that the value of β_1 differs significantly from zero.

Simple Linear Regression

EXAMPLE 13.10 Refer to Example 13.9. Construct a 99% confidence interval for β_1.

Solution   The general form of a 99% confidence interval for β_1 is

$$\hat{\beta}_1 \pm t_{.005} s_{\hat{\beta}_1}$$

where $t_{.005} = 5.841$ is based on $n - 2 = 3$ degrees of freedom, and

$$s_{\hat{\beta}_1} = s/\sqrt{SS_{xx}} = 2.13/\sqrt{49{,}684.8} = .00956$$

Now, the desired confidence interval is:

$$-.043 \pm 5.841(.00956) = -.043 \pm .0558, \text{ or } (-.0988, .0128)$$

Note that the interval contains the value zero, which reflects a lack of sufficient evidence that x contributes information for the prediction of y. This is consistent with the result of the previous example, in which we failed to reject H_0: $\beta_1 = 0$.

EXERCISES

13.5 Refer to Exercises 13.2 and 13.3. Test the null hypothesis that x contributes no information for the prediction of y against the alternative hypothesis that these variables are linearly related with positive slope, i.e., that GMAT scores tend to increase as GPA increases. Use $\alpha = .025$.

13.6 Refer to Exercises 13.2 and 13.3. Construct a 95% confidence interval for β_1, the mean increase in GMAT score for a unit increase in GPA.

13.7 Advertising constitutes a large portion of many companies' budgets. To determine if there is a linear relationship between the amount of money spent on advertising and the amount of sales, ten drug and cosmetic companies were surveyed, with the following results:

Company	Advertising Expenditure, x (millions of dollars)	Sales, y (millions of dollars)
1	115.0	955.0
2	72.0	870.5
3	62.0	182.6
4	44.5	880.2
5	30.0	438.6
6	27.5	128.8
7	20.3	68.4
8	16.8	769.4
9	48.0	231.8
10	83.9	490.6

For these data,

$$SS_{xx} = 8,887.04 \quad \bar{x} = 52.00$$
$$SS_{yy} = 1,061,998.689 \quad \bar{y} = 501.59$$
$$SS_{xy} = 45,466.48$$

Do the data provide sufficient evidence to indicate that advertising expenditures contribute information for predicting sales? Use $\alpha = .05$.

13.6 Correlation: A Measure of the Usefulness of the Model
13.7 The Coefficient of Determination

EXAMPLE 13.11 Refer to Examples 13.2–13.5.

    a. Compute the coefficient of correlation, r, for the alumnus donation example. Interpret its value.

    b. Compute the coefficient of determination, r^2, for the sample data. Interpret its value.

Solution

    a. The coefficient of correlation, r, is computed as follows:

$$r = \frac{SS_{xy}}{\sqrt{SS_{xx}SS_{yy}}}$$

where we previously obtained the values $SS_{xy} = 2,777$ and $SS_{xx} = 995.6$.

Now

$$SS_{yy} = \sum y^2 - \frac{(\sum y)^2}{n} = (25^2 + 20^2 + \cdots + 50^2 + 10^2) - \frac{(670)^2}{15}$$
$$= 51,100 - 29,926.667 = 21,173.333$$

Thus,

$$r = \frac{2,777}{\sqrt{(995.6)(21,173.333)}} = .605$$

The moderately large, positive value of r implies a linear relationship between y and x, in which y tends to increase as x increases.

    b. The coefficient of determination is

$$r^2 = (.605)^2 = .366$$

Note that this value may also be obtained as follows:

$$r^2 = \frac{SS_{yy} - SSE}{SS_{yy}} = \frac{21,173.333 - 13,427.53}{21,173.333} = .366$$

Simple Linear Regression

This represents the proportion of the sum of squares of the deviations of the sample y values about their mean that may be attributed to a linear relation between y and x. In other words, there is a 36.6% reduction in the sum of squared prediction errors when the least squares line $\hat{y} = \hat{\beta}_0 + \hat{\beta}_1 x$, instead of $\bar{y}$, is used to predict y.

EXAMPLE 13.12 A radio station has experimented with contests which give away money and prizes in an attempt to increase the size of its listening audience, as measured by a listener rating service. Results from the last two weeks are shown in the following table.

Value of Prizes, x (hundreds of dollars)	Listener Rating, y
0	11.1
3	11.5
7	14.0
0	12.0
1	9.0
6	11.0
20	23.0
10	22.4
5	19.3
10	18.6
12	18.5
20	21.3
0	17.8
12	15.7

For these data, the following values have been computed:

$SS_{xx} = 605.428, \bar{x} = 7.57$
$SS_{yy} = 283.437, \bar{y} = 16.09$
$SS_{xy} = 305.314$

a. Obtain the least squares prediction equation.

b. Compute the value of r, the coefficient of correlation.

c. Do the data provide sufficient evidence to indicate a nonzero population correlation between x and y? Use $\alpha = .05$.

Solution a. The slope of the least squares line is

$$\hat{\beta}_1 = \frac{SS_{xy}}{SS_{xx}} = \frac{305.314}{605.428} = .50$$

and the y-intercept is

$$\hat{\beta}_0 = \bar{y} - \hat{\beta}_1 \bar{x} = 16.09 - .50(7.57) = 12.31$$

Thus, the least squares prediction equation is

$$\hat{y} = 12.31 + .50x$$

b. $r = \dfrac{SS_{xy}}{\sqrt{SS_{xx}SS_{yy}}} = \dfrac{305.314}{\sqrt{(605.428)(283.437)}} = .74$

c. The test of

H_0: Population coefficient of correlation $\rho = 0$

against

H_a: $\rho \ne 0$

is equivalent to the test of

H_0: $\beta_1 = 0$

against

H_a: $\beta_1 \ne 0$

At $\alpha = .05$, the null hypothesis will be rejected if

$t < -t_{.025}$ or $t > t_{.025}$

i.e.,

$t < -2.179$ or $t > 2.179$

where the distribution of t has $n - 2 = 12$ degrees of freedom.

It is necessary to compute the value of s:

$$s = \sqrt{\dfrac{SSE}{n-2}} = \sqrt{\dfrac{SS_{yy} - \hat{\beta}_1 SS_{xy}}{n-2}} = \sqrt{\dfrac{283.437 - .50(305.314)}{12}}$$

$$= \sqrt{10.8983} = 3.30$$

Then the statistic is

$$t = \dfrac{\hat{\beta}_1 - 0}{s_{\hat{\beta}_1}} = \dfrac{\hat{\beta}_1}{s/\sqrt{SS_{xx}}} = \dfrac{.50}{3.30/\sqrt{605.428}} = 3.73$$

Since the value of t falls within the rejection region, we reject H_0 and conclude that x does contribute information for the prediction of y. In other words, there is sufficient evidence in the sample to conclude that the population coefficient of correlation between x and y is significantly nonzero.

Simple Linear Regression

EXERCISES

13.8 Refer to Exercises 13.2 and 13.3.

  a. Compute the coefficient of correlation, r, between GMAT score and GPA for this sample. Interpret its value.

  b. Compute the coefficient of determination, r^2, for the data. Interpret its value.

13.9 Refer to Exercise 13.8. Do the data provide sufficient evidence to conclude that the population coefficient of correlation between x and y is significantly greater than zero? Use $\alpha = .01$.

13.10 Refer to Exercise 13.7. Determine the coefficient of determination and explain its significance in terms of the problem.

13.8 Using the Model for Estimation and Prediction

EXAMPLE 13.13 For the alumnus donation data of Example 13.2, find a 90% confidence interval for the mean amount of money donated per year when the number of years out of school is 20.

Solution A 90% confidence interval for the mean value of y is

$$\hat{y} \pm t_{.05} s \sqrt{\frac{1}{n} + \frac{(x - \bar{x})^2}{SS_{xx}}}$$

where $t_{.05} = 1.771$ is based on $n - 2 = 13$ degrees of freedom.

Recall that $s = 32.14$, $\bar{x} = 15.4$, $n = 15$, and $SS_{xx} = 995.6$ for the alumnus donation example. Now, the estimated value of y when $x = 20$ is

$$\hat{y} = \hat{\beta}_0 + \hat{\beta}_1(20) = 1.716 + 2.789(20) = 57.496$$

Substitution into the general formula for the confidence interval yields:

$$57.496 \pm 1.771(32.14) \sqrt{\frac{1}{15} + \frac{(20 - 15.4)^2}{995.6}}$$

$$= 57.496 \pm 16.878, \text{ or } (40.618, 74.374)$$

We estimate, with 90% confidence, that the interval from $40.62 to $74.37 encloses the mean amount donated per year to the school when the number of years out of school is 20.

EXAMPLE 13.14 Predict the amount to be donated per year to the school by a particular alumnus who has been out of school for 20 years. Use a 90% prediction interval.

Solution To predict the amount to be donated by a particular alumnus for whom $x = 20$, we calculate the 90% prediction interval as

$$\hat{y} \pm t_{.05} s \sqrt{1 + \frac{1}{n} + \frac{(x - \bar{x})^2}{SS_{xx}}}$$

where $t_{.05} = 1.771$ (13 df), $s = 32.14$, $n = 15$, $\bar{x} = 15.4$, $SS_{xx} = 995.6$, and

$$\hat{y} = \hat{\beta}_0 + \hat{\beta}_1(20) = 1.716 + 2.789(20) = 57.496$$

Thus, the desired prediction interval is

$$57.496 \pm 1.771(32.14) \sqrt{1 + \frac{1}{15} + \frac{(20 - 15.4)^2}{995.6}}$$

$$= 57.496 \pm 59.369, \text{ or } (-1.873, 116.865)$$

This very wide prediction interval contains negative values, which have no meaning in the context of this problem. We predict that the alumnus will not donate more than $116.87 per year to the college.

EXAMPLE 13.15 Refer to Example 13.12. Predict the listener rating for tomorrow if the value of prizes to be given away is $1,500. Use a 95% prediction interval.

Solution To predict the listener rating for a particular day on which $x = 15$, we compute the 95% prediction interval as

$$\hat{y} \pm t_{.025} s \sqrt{1 + \frac{1}{n} + \frac{(x - \bar{x})^2}{SS_{xx}}}$$

where $t_{.025} = 2.179$ (12 df), $s = 3.30$, $n = 14$, $\bar{x} = 7.57$, $SS_{xx} = 605.428$, and

$$\hat{y} = \hat{\beta}_0 + \hat{\beta}_1(15) = 12.31 + .50(15) = 19.81$$

Thus, the prediction interval is

$$19.81 \pm 2.179(3.30) \sqrt{1 + \frac{1}{14} + \frac{(15 - 7.57)^2}{605.428}}$$

$$= 19.81 \pm 7.75, \text{ or } (12.06, 27.56)$$

We predict that tomorrow's listener rating will fall in the interval from 12.06 to 27.56.

EXERCISES

13.11 Refer to Exercise 13.2. Construct a 90% confidence interval for the mean GMAT score when the GPA is 3.6.

Simple Linear Regression

13.12 Predict the GMAT score for a student whose GPA is 3.6. Use a 90% prediction interval.

13.13 Compare the lengths of the intervals obtained in Exercises 13.11 and 13.12.

13.9 Simple Linear Regression: An Example

EXAMPLE 13.16 A real estate broker has been collecting data on home sales so that he may investigate the relationship between the value of a purchased home, y, and the annual family income of the buyer, x. Data from six recent sales are shown in the following table:

Annual Family Income x, (thousands of dollars)	Value of Home, y (thousands of dollars)
45.2	83.8
47.4	98.9
52.0	99.5
54.6	111.0
59.8	113.8
68.0	142.5

a. Develop a simple linear model for the relationship between y and x.

b. Use the data in the table to obtain the least squares prediction equation.

c. Do the data provide sufficient evidence to indicate that the annual family income of the buyer contributes information about the value of a home purchased? Use a significance level of $\alpha = .05$.

d. Calculate and interpret the value of r^2 for these data.

e. If the annual family income of a particular buyer is $53,000, form a 95% prediction interval for the value of a home to be purchased.

f. Find a 95% confidence interval for the mean value of a home purchase when the annual family income is $57,500.

Solution

a. We hypothesize the following straight-line probabilistic model:

$$y = \beta_0 + \beta_1 x + \epsilon$$

where y = value of the purchased home (in thousands of dollars) and x = annual family income of buyer (in thousands of dollars).

b. We perform the following preliminary calculations:

$$SS_{xx} = 353.50, \quad \bar{x} = 54.50$$
$$SS_{yy} = 1{,}973.215 \quad \bar{y} = 108.25$$
$$SS_{xy} = 807.71$$

Now, the least squares line has slope

$$\hat{\beta}_1 = \frac{SS_{xy}}{SS_{xx}} = \frac{807.71}{353.50} = 2.28$$

and y-intercept

$$\hat{\beta}_0 = \bar{y} - \hat{\beta}_1 \bar{x} = -16.28$$

Thus, the least squares prediction equation is

$$\hat{y} = -16.28 + 2.28x$$

c. It is necessary to perform a test of the null hypothesis

$$H_0: \beta_1 = 0$$

against the alternative hypothesis

$$H_a: \beta_1 \neq 0$$

At significance level $\alpha = .05$, we will reject the null hypothesis if

$$t < -t_{.025} \text{ or } t > t_{.025}$$

i.e.,

$$t < -2.776 \text{ or } t > 2.776$$

where t has $n - 2 = 4$ degrees of freedom.

The test statistic is

$$t = \frac{\hat{\beta}_1 - 0}{s_{\hat{\beta}_1}} = \frac{\hat{\beta}_1}{s/\sqrt{SS_{xx}}}$$

where

$$s = \sqrt{\frac{SSE}{n-2}} = \sqrt{\frac{SS_{yy} - \hat{\beta}_1 SS_{xy}}{n-2}} = \sqrt{\frac{1{,}973.215 - 2.28(807.71)}{4}}$$

$$= \sqrt{32.90905} = 5.74$$

Thus,

$$t = \frac{2.28}{5.74/\sqrt{353.50}} = 7.47$$

The computed value of the test statistic lies within the rejection region. We thus conclude that the annual family income of the buyer contributes information for the prediction of the value of a home purchased.

Simple Linear Regression

(Note that this test procedure requires the assumptions that ϵ is normally distributed with mean 0, constant variance σ^2, and that the errors are independent.)

d. The coefficient of determination is

$$r^2 = \frac{SS_{yy} - SSE}{SS_{yy}}$$

where

$$SSE = SS_{yy} - \hat{\beta}_1 SS_{xy} = 1{,}973.215 - 2.28(807.71) = 131.6362$$

Then

$$r^2 = \frac{1{,}973.215 - 131.6362}{1{,}973.215} = .933$$

There is a 93.3% reduction in the sum of squared prediction errors when the least squares line $\hat{y} = -.58 + 2.28x$, instead of $\bar{y} = 78.25$, is used to predict y.

e. To predict the value of a home to be purchased by a particular family for whom $x = 53$, we calculate the 95% prediction interval:

$$\hat{y} \pm t_{.025} s \sqrt{1 + \frac{1}{n} + \frac{(x - \bar{x})^2}{SS_{xx}}}$$

where $t_{.025} = 2.776$ (4 df) and

$$\hat{y} = -16.28 + 2.28(53) = 104.56$$

Thus, the prediction interval is

$$104.56 \pm 2.776(5.74)\sqrt{1 + \frac{1}{6} + \frac{(53 - 54.50)^2}{353.50}}$$

$$= 104.56 \pm 17.26, \text{ or } (87.3, 121.82)$$

We predict, with 95% confidence, that the value of the home purchased by this family will fall in the interval from $87,300 to $121,820.

f. A 95% confidence interval for the mean value of a home purchase when the annual family income is $57,500 (i.e., $x = 57.5$) is computed as

$$\hat{y} \pm t_{.025} s \sqrt{\frac{1}{n} + \frac{(x - \bar{x})^2}{SS_{xx}}}$$

where $\hat{y} = -16.28 + 2.28(57.5) = 114.82$

The desired confidence interval is then

$$114.82 \pm 2.776(5.74)\sqrt{\frac{1}{6} + \frac{(57.5 - 54.5)^2}{353.50}}$$

$$= 114.82 \pm 6.98, \text{ or } (107.84, 121.8)$$

We estimate, with 95% confidence, that the interval from $107,840 to $121,800 encloses the mean value of a home purchase when the annual family income is $57,500.

EXERCISE

13.14 In order to introduce consumers to a product, manufacturers often advertise with coupons which discount the store price of an item. Marketing executives for a frozen food processor have experimented to determine the relationship between the value of a coupon and the number of coupons that will be used by consumers within one month. The following recent data have been collected:

Coupon Value, x (cents)	Number of Coupons Used, y (hundreds)
5	13.28
10	11.03
12	12.92
15	15.03
25	18.46
50	22.15

a. Hypothesize a simple linear model for the relationship between y and x.

b. Obtain the least squares prediction equation.

c. Test the null hypothesis (at $\alpha = .05$) that x contributes no information for the prediction of y against the alternative that the values of y tend to increase as x increases. State any assumptions required for the validity of the test procedure.

d. Calculate the values of r and r^2 for the data. Interpret their values.

e. Construct a 95% confidence interval for the mean number of coupons used when the value of the coupon is 20 cents.

Simple Linear Regression

CHAPTER FOURTEEN

Multiple Regression

Summary

This chapter discussed the steps to be followed in a **multiple regression analysis**, a procedure for modeling a dependent variable y as a function of k independent variables, $x_1, x_2, \ldots, x_k$. The methodology is much the same as for simple straight-line models:

1. A probabilistic model is hypothesized.

2. The unknown parameters in the deterministic component of the model are estimated using the method of least squares.

3. The probability distribution of the random error component ϵ is specified.

4. Inferences are performed to assess the usefulness of the model. The regression assumptions are checked by residual analysis.

5. If the model is judged to be satisfactory, it may be used for estimation and for prediction of y values to be observed in the future.

14.1 Multiple Regression: The Model and the Procedure
14.2 Fitting the Model: The Least Squares Approach

EXAMPLE 14.1 Most power companies now permit their residential customers to pay their electricity bills in equal monthly payments, rather than on the actual amount of electricity used each month. This reduces the high winter bills and increases the low summer bills.

A local electric company has proposed the following model to predict a customer's bill for next year:

$$y = \beta_0 + \beta_1 x_1 + \beta_2 x_2 + \epsilon$$

where y = total amount to be billed next year, x_1 = total amount billed in current year, and x_2 = inflation rate in current year.

Records from previous years yielded the following data for a set of 20 randomly selected residential customers:

Amount Billed in Following Year, y ($)	Amount Billed in Prior Year, x_1 ($)	Inflation Rate in Prior Year, x_2 (%)
405	355	6.2
285	240	3.6
607	500	10.3
468	465	4.6
627	550	11.4
572	487	12.6
503	498	10.3
385	390	4.6
435	428	3.6
410	401	7.9
393	415	8.4
630	594	12.6
473	468	8.4
355	315	4.6
453	404	11.4
485	468	9.8
304	255	6.8
563	510	10.3
511	476	6.2
343	328	7.9

A portion of the output from the Statistical Analysis System (SAS) multiple regression routine for these data is shown below:

SOURCE	DF	SUM OF SQUARES	MEAN SQUARE	F VALUE	PR > F
MODEL	2	186699.0116	93349.51	98.29	0.0001
ERROR	17	16145.5384	949.74		
CORRECTED TOTAL	19	202844.5500		R-SQUARE	ROOT MSE
				.920	30.8178

PARAMETER	ESTIMATE	T FOR H0: PARAMETER=0	PR > |T|	STD ERROR OF ESTIMATE
INTERCEPT	16.5901	0.50	0.6233	33.16066
X1	0.9197	9.17	0.0001	0.10034
X2	6.2820	1.98	0.0640	3.17090

a. Identify the least squares prediction equation.

b. Predict a customer's bill for next year if he was billed $450 this year and the inflation rate is currently 11.8%.

Solution   a. The least squares estimates of β_0, β_1, and β_2 appear in the column labeled ESTIMATE:

$$\hat{\beta}_0 = 16.59, \hat{\beta}_1 = .920, \text{ and } \hat{\beta}_2 = 6.282$$

Multiple Regression

Thus, the least squares prediction equation is:

$$\hat{y} = 16.59 + .920x_1 + 6.282x_2$$

b. For a customer with $x_1 = 450$ and $x_2 = 11.8$, the amount predicted for next year's bill is:

$$\hat{y} = 16.59 + .920(450) + 6.282(11.8) = 504.7$$

We predict a bill of $505 for this customer next year. (A measure of the reliability of such predictions is discussed in Section 14.5.)

EXERCISE

14.1 Admission criteria at graduate schools of business are often based on a student's undergraduate grade point average (GPA) and the score on the Graduate Management Aptitude Test (GMAT). Recently, questions have been raised about the usefulness of these two variables in the prediction of a student's GPA in graduate school. Data from nine recent graduates with an MBA degree are shown below:

GPA in Graduate School y	Undergraduate GPA x_1	GMAT Score x_2
3.83	3.90	680
3.72	3.89	660
3.61	3.35	710
3.50	3.20	530
3.46	3.78	490
3.41	3.90	520
3.30	3.10	560
3.26	2.93	610
3.07	3.21	550

The model

$$y = \beta_0 + \beta_1 x_1 + \beta_2 x_2 + \epsilon$$

was fit to the data; results from the SAS routine are shown below (values have been rounded):

SOURCE	DF	SUM OF SQUARES	MEAN SQUARE	F VALUE	PR > F
MODEL	2	.323	.161	7.68	.022
ERROR	6	.126	.021		
CORRECTED TOTAL	8	.449		R-SQUARE	ROOT MSE
				.7194	.1449

PARAMETER	ESTIMATE	T FOR H0: PARAMETER=0	PR > |T|	STD ERROR OF ESTIMATE
INTERCEPT	1.2093	2.10	.081	.5771
X1	.3710	2.82	.030	.1310
X2	.0021	2.50	.048	.00084

a. Identify the least squares prediction equation.

b. What would be the predicted graduate school GPA for a student with an undergraduate GPA of 3.65 and a GMAT score of 640?

14.3 Model Assumptions

EXAMPLE 14.2 State the assumptions made about the probability distribution of ϵ in a multiple regression model.

Solution The random error component ϵ is assumed to have a normal distribution with mean zero and a variance σ^2 which is constant for all configurations of the independent variables $x_1, x_2, \ldots, x_k$. In addition, the errors associated with different observations of y are assumed to be independent.

EXAMPLE 14.3 Refer to Example 14.1.

a. Obtain the value of SSE for the electricity bills example.

b. Estimate the variance of ϵ in the model.

Solution a. The value for SSE appears in the printout as the SUM OF SQUARES for ERROR. Thus,

$$SSE = 16,145.5384$$

b. The estimator of σ^2, the variance of ϵ, is given by the mean square for error:

$$MSE = \frac{SSE}{n - (\text{number of estimated } \beta \text{ parameters})}$$

In our example, $n = 20$ and three β parameters are estimated from the data. Thus,

$$MSE = \frac{16,145.5384}{20 - 3} = 949.74$$

Note that this value appears in the printout as the MEAN SQUARE for ERROR.

EXERCISE

14.2 Refer to Exercise 14.1.

a. Obtain the value of SSE from the printout for the graduate school GPA data.

b. Compute MSE, the estimate of σ^2 for the model.

Multiple Regression

14.4 Estimating and Testing Hypotheses About the β Parameters

EXAMPLE 14.4 Refer to Examples 14.1–14.3. Is there evidence to indicate that the following year's bill increases as the current year's inflation rate increases? Use a significance level of $\alpha = .05$.

Solution The hypothesis of interest concerns the parameter β_2:

$$H_0: \beta_2 = 0$$
$$H_a: \beta_2 > 0$$

where β_2 is the mean increase in the following year's bill for a 1% increase in the current inflation rate, when the amount billed in the current year is held constant.

For $\alpha = .05$, the rejection region consists of values of t such that

$$t > t_{.05} = 1.740$$

where the distribution of t has degrees of freedom equal to

$$n - (\text{number of } \beta \text{ parameters estimated in model}) = 20 - 3 = 17$$

(Note from Example 14.3 that this is exactly the number of degrees of freedom associated with the estimate of σ^2 for this model.)

The test statistic is

$$t = \frac{\hat{\beta}_2}{s_{\hat{\beta}_2}}$$

where $s_{\hat{\beta}_2}$, the estimated standard deviation of the model coefficient $\hat{\beta}_2$, is found in the printout column labeled STD ERROR OF ESTIMATE. Then, for our example,

$$t = \frac{6.2820}{3.1709} = 1.98$$

This computed value lies within the rejection region. We thus conclude that the mean amount of the following year's bill increases as the current inflation rate increases.

It should be noted that the t statistic for testing the null hypothesis that an individual β parameter is equal to zero is shown in the column of the printout labeled T FOR H0: PARAMETER=0. The value in the column headed PR > |T| represents the two-tailed probability that

$$t < -t_{\text{computed}} \quad \text{or} \quad t > t_{\text{computed}}$$

In our example,

$$P(t < -1.98 \text{ or } t > 1.98) = .0640$$

and thus,

$$P(t > 1.98) = \frac{.0640}{2} = .0320$$

This implies that we would reject the null hypothesis in favor of our one-sided alternative for any value of α greater than .0320.

Note that the validity of this test procedure depends upon the following assumptions about the probability distribution of ϵ: the distribution of ϵ is normal with mean 0 and variance σ^2, which is constant for all values of x, and the errors are independent.

EXAMPLE 14.5 An appliance store has hypothesized the following quadratic model relating the number of air-conditioning units sold daily, y, to the daily high temperature, x:

$$y = \beta_0 + \beta_1 x + \beta_2 x^2 + \epsilon$$

Data collected from a random sample of 40 summer days were fit to the model, with the following results:

SOURCE	DF	SUM OF SQUARES
MODEL	2	9.2
ERROR	37	14.8
CORRECTED TOTAL	39	24.0

PARAMETER	ESTIMATE	STD ERROR OF ESTIMATE
INTERCEPT	-9.63	.371
X	0.19	.048
X * X	0.01	.0017

a. Give an interpretation of β_2 in the context of the problem.

b. Test to determine if the quadratic term makes a significant contribution to the model, at $\alpha = .01$.

Solution

a. The parameter β_2, the coefficient of x^2, measures the curvature in the response curve relating y and x. The appliance store may believe, for example, that the number of air-conditioning units sold increases almost linearly as the daily high temperature increases through the lower range of temperatures. Then, in the upper range of daily temperatures, the increase in the number of units sold for a unit degree increase in temperature may begin to increase.

b. The parameter of interest is β_2, and the appliance store may wish to test its hypothesis, stated in part a:

$H_0: \beta_2 = 0$ (Response curve is linear through the entire range of temperatures.)

$H_a: \beta_2 > 0$ (Upward curvature exists in the response curve.)

At $\alpha = .01$, the null hypothesis is rejected if

$$t > t_{.01} \approx z_{.01} = 2.33$$

where t has $n - 3 = 37$ degrees of freedom, and thus is very similar to the standard normal (z) distribution.

From the computer printout, we obtain $\hat{\beta}_2 = .01$ and $s_{\hat{\beta}_2} = .0017$. Thus, the computed value of the test statistic is

$$t = \frac{\hat{\beta}}{s_{\hat{\beta}_2}} = \frac{.01}{.0017} = 5.88$$

The computed t value is within the rejection region and we conclude that there is significant upward curvature in the response curve relating y and x.

(This test, as with all inferential procedures in a multiple regression context, requires the assumptions about the probability distribution of ϵ as stated in the previous example.)

EXERCISES

14.3   Refer to Exercises 14.1 and 14.2.

   a. Is there sufficient evidence (at $\alpha = .01$) that the mean graduate school GPA increases as undergraduate GPA increases, when the value of GMAT score is held constant?

   b. State any assumptions required for the validity of the test procedure used in part a.

14.4   Data on ten patients in a psychiatric hospital were collected for the following variables:

   y = hostility score (measured by a standardized test)
   x_1 = age
   x_2 = number of prior hospitalizations for psychiatric treatment

   The following least squares equation was obtained:

   $$\hat{y} = 20 + .1x_1 + 1.2x_2$$

   with

   $s_{\hat{\beta}_1} = .064$ and $s_{\hat{\beta}_2} = .45$

   a. State the probabilistic model which was fit to the data.

   b. Use these results to determine whether there is evidence (at $\alpha = .05$) to indicate that the mean hostility score depends on the number of prior hospitalizations for psychiatric treatment.

14.5 Checking the Usefulness of a Model: R^2 and the Analysis of Variance F-Test

EXAMPLE 14.6 Refer to Examples 14.1–14.3.

a. Obtain the value of R^2, the multiple coefficient of determination, for the electricity bills data. Interpret its value.

b. Is there evidence to indicate that the overall model is useful for predicting the following year's bill? Use $\alpha = .01$.

Solution

a. The SAS multiple regression routine gives the multiple coefficient of determination, R-SQUARE, in the output. For our example, $R^2 = .920$. Thus, 92% of the sample variation of the y values is accounted for by the regression model. In other words, the sum of squared prediction errors is reduced by 92% when the least squares prediction equation $\hat{y} = 16.59 + .92x_1 + 6.282x_2$ is used, instead of $\bar{y}$, to predict the value of y.

b. To test the global utility of the model, we test

$$H_0: \beta_1 = \beta_2 = 0$$

against

H_a: At least one of the coefficients differs from zero

The test, which requires the standard assumptions about the probability distribution of ϵ, is based on an F statistic with k degrees of freedom in the numerator and $[n - (k + 1)]$ degrees of freedom in the denominator, where k is the number of independent variables in the model.

For our example, $n = 20$, $k = 2$, and the null hypothesis will be rejected at $\alpha = .01$ if

$$F > F_{.01} = 6.11$$

where F is based on 2 numerator degrees of freedom and 17 denominator degrees of freedom.

The test statistic is

$$F = \frac{R^2/k}{(1 - R^2)/[n - (k + 1)]} = \frac{.920/2}{(1 - .920)/[20 - (2 + 1)]} = 97.75$$

This value exceeds the tabulated critical value of 6.11; thus, we conclude that at least one of the model coefficients β_1 and β_2 is significantly different from zero, and that the model

$$y = \beta_0 + \beta_1 x_1 + \beta_2 x_2 + \epsilon$$

is useful for predicting the following year's electricity bill.

(Note that the computer routine provides the computed F for the test of the overall utility of the model. In the printout given in Example 14.1, we find $F = 98.29$; this differs from our computed value only because of rounding errors.)

EXAMPLE 14.7 Refer to Example 14.5.

a. Use the information given to compute the value of R^2 for the air-conditioning sales data.

b. Conduct the global F test of model utility at the $\alpha = .05$ level of significance.

Solution a. The multiple coefficient of determination is defined as

$$R^2 = 1 - \frac{SSE}{SS_{yy}}$$

where SSE is referred to as SUM OF SQUARES for ERROR in the printout, and SS_{yy} the CORRECTED TOTAL SUM OF SQUARES. For the air-conditioning sales data,

$$R^2 = 1 - \frac{14.8}{24.0} = 1 - .617 = .383$$

b. The elements of the test are

$H_0: \beta_1 = \beta_2 = 0$
H_a: At least one of the coefficient is nonzero.

The test statistic is based on $k = 2$ numerator degrees of freedom and $n - (k + 1) = 40 - 3 = 37$ denominator degrees of freedom; thus, we will reject the null hypothesis at $\alpha = .05$ if

$$F > F_{.05} \approx 3.26$$

The computed value of the test statistic is:

$$F = \frac{R^2/k}{(1 - R^2)/[n - (k + 1)]} = \frac{.383/2}{(1 - .383)/37} = 11.48$$

We reject the null hypothesis and conclude that the model is useful for predicting air-conditioning sales.

EXERCISES

14.5 Refer to Exercises 14.1–14.3. Compute the value of the multiple coefficient of determination, R^2, for the data on graduate school GPA. Interpret its value.

14.6 Refer to Exercises 14.1–14.3. Perform the global F test of the utility of the model. Use a significance level of $\alpha = .05$.

14.6 Using the Model for Estimation and Prediction

EXAMPLE 14.8 Refer to Example 14.1b, where we used the least squares equation to predict the amount of next year's electricity bill for a particular customer with $x_1 = 450$ and $x_2 = 11.8$. Suppose the corresponding 95% prediction interval were (253.9, 755.5). Would you expect a 95% confidence interval for $E(y)$, the mean electricity bill for all customers with $x_1 = 450$ and $x_2 = 11.8$ to be narrower or wider than this interval? Explain.

Solution As was the case with the simple linear regression procedures of Chapter 13, the least squares equation yields the same value for both $E(y)$ and for the prediction of a future value of y for a specific configuration of values of the independent variables. (In this example, with $x_1 = 450$ and $x_2 = 11.8$, this common value was $\hat{y} = 504.7$, obtained in Example 14.1.) Furthermore, the confidence interval for the mean value will be narrower than the prediction interval for a particular value of y, because of the additional variability of ϵ in predicting a particular value of y.

EXERCISE

14.7 Refer to Exercises 14.1–14.3. Suppose that the 90% confidence interval for the mean graduate school GPA when $x_1 = 3.65$ and $x_2 = 640$ is given by (3.822, 3.992). Interpret this interval.

14.7 Multiple Regression: An Example

EXAMPLE 14.9 A college administrator would like to be able to predict the number of credit hours students will take in a given quarter, so that he can plan the budget accordingly. He believes several factors influence the number of hours taken by each student, and proposed the following model to predict the number of credit hours taken:

$$y = \beta_0 + \beta_1 x_1 + \beta_2 x_2 + \beta_3 x_3 + \epsilon$$

where
y = number of credit hours taken

x_1 = total number of credit hours previously taken

$x_2 = \begin{cases} 1 \text{ if student is in upper division} \\ 0 \text{ if student is in lower division} \end{cases}$

and

$x_3 = \begin{cases} 1 \text{ if fall or winter quarter} \\ 0 \text{ if summer or spring quarter} \end{cases}$

A random sample of 20 students produced the following data.

Number of Credit Hours y	Number of Credit Hours Previously Taken x_1	Dummy Variable for Division x_2	Dummy Variable for Quarter x_3
16	95	1	1
19	142	1	1
14	128	1	1
12	165	1	1
15	131	1	1
13	104	1	0
15	153	1	0
13	128	1	0
12	139	1	0
14	101	1	0
19	15	0	1
18	73	0	1
16	31	0	1
17	65	0	1
15	93	0	1
13	26	0	0
15	100	0	0
16	38	0	0
17	85	0	0
16	70	0	0

A portion of the SAS multiple regression analysis of these data is shown below:

SOURCE	DF	SUM OF SQUARES	MEAN SQUARE	F VALUE	PR > F
MODEL	3	33.0755	11.025	3.48	.0407
ERROR	16	50.6745	3.167		
CORRECTED TOTAL	19	83.7500		R-SQUARE	ROOT MSE
				.395	1.780

PARAMETER	ESTIMATE	T FOR H0: PARAMETER=0	PR > |T|	STD ERROR OF ESTIMATE
INTERCEPT	15.751	13.50	0.0001	1.167
X1	-0.00670	-0.43	0.6756	0.0157
X2	-1.438	-1.07	0.3010	1.345
X3	1.696	2.13	0.0490	0.796

a. Identify the least squares model that was fit to the data.

b. What is the predicted number of credit hours an upper division student will take in spring quarter if he has already taken 100 credit hours?

c. Specify the probability distribution of ϵ.

d. Obtain an estimate of σ^2, the variance of ϵ.

e. Obtain the value of R^2 and interpret it.

f. Conduct the global F test of model utility. Use a significance level of .05.

g. Is there evidence that the number of credit hours taken tends to decrease as the number of credit hours previously taken increases? Use $\alpha = .05$.

Solution

a. The least squares model is:

$$\hat{y} = 15.751 - .00670x_1 - 1.438x_2 + 1.696x_3$$

b. To predict the number of credit hours to be taken in spring quarter by an upper division student who has already taken 100 credit hours, we substitute $x_1 = 100$, $x_2 = 1$, and $x_3 = 0$ into the prediction equation:

$$\hat{y} = 15.751 - .00670(100) - 1.438(1) + 1.696(0) = 13.6$$

c. We assume that the probability distribution of ϵ is normal, with a mean of zero and a constant variance σ^2. In addition, the errors are assumed to be independent.

d. The estimate of σ^2, the variance of the random error component ϵ, is given in the SAS printout as MEAN SQUARE for ERROR:

$$\text{MSE} = 3.167$$

Note that this quantity may be computed directly as

$$\text{MSE} = \frac{\text{SSE}}{n - (k + 1)} = \frac{50.6745}{20 - (3 + 1)} = 3.167$$

e. The value of the multiple coefficient of determination is shown in the printout:

$$R^2 = .395$$

Note that this quantity may also be computed as

$$R^2 = 1 - \frac{\text{SSE}}{\text{SS}_{yy}} = 1 - \frac{50.6745}{83.7500} = 1 - .605 = .395$$

Thus, only 39.5% of the sample variation in the y values is accounted for by the least squares model.

f. The elements of the test are:

H_0: $\beta_1 = \beta_2 = \beta_3 = 0$
H_a: At least one of the coefficients is nonzero

Multiple Regression

The test is based on an F statistic with $k = 3$ numerator degrees of freedom and $n - (k + 1) = 20 - 4 = 16$ denominator degrees of freedom. For $\alpha = .05$, the null hypothesis will be rejected for

$$F > F_{.05} = 3.24$$

The test statistic is

$$F = \frac{R^2/k}{(1 - R^2)/[n - (k + 1)]} = \frac{.395/3}{(1 - .395)/16} = 3.48$$

(This value may be obtained directly from the printout.)

We reject the null hypothesis and conclude, at significance level $\alpha = .05$, that the model is useful for predicting number of credit hours to be taken.

g. The parameter of interest is β_1, and we wish to test

$$H_0: \beta_1 = 0$$

against

$$H_a: \beta_1 < 0$$

At significance level $\alpha = .05$, we will reject H_0 if

$$t < -t_{.05} = -1.746$$

where t is based on $n - (k + 1) = 16$ degrees of freedom.

The test statistic is computed as

$$t = \frac{\hat{\beta}_1}{s_{\hat{\beta}_1}} = \frac{-.00670}{.0157} = -.43$$

This value, which may be obtained directly from the printout, does not lie within the rejection region. There is insufficient evidence to conclude that the number of credit hours taken decreases as the number of hours previously taken increases when the dummy variables for division and quarter are held constant.

14.9 Residual Analysis: Checking the Regression Assumptions

EXAMPLE 14.10 Refer to Example 14.1. The observed y values, the predicted y values, $\hat{y}$, and the corresponding residuals are shown below. The residuals are then plotted on the vertical axis against the independent variable x_1, the amount billed in the current year, and against the independent variable x_2, the inflation rate in the current year.

a. Verify that each residual is equal to the difference between the observed y value and the estimated mean value $\hat{y}$.

b. Analyze the residual plots.

Obs.	Actual y ($)	Predicted y ($)	Residual
1	405	382.0312	22.96884
2	285	259.9327	25.06726
3	607	541.1435	65.85651
4	468	473.1467	−5.146667
5	627	594.0385	32.96149
6	572	543.6360	28.36395
7	503	539.3041	−36.30408
8	385	404.1693	−19.16934
9	435	432.8358	2.164154
10	410	435.0166	−25.01663
11	393	451.0334	−58.03339
12	630	642.0437	−12.04370
13	473	499.7774	−26.77737
14	355	335.1920	19.80795
15	453	459.7627	−6.762726
16	485	508.5722	−23.57217
17	304	293.8306	10.16940
18	563	550.3405	12.65955
19	511	493.3145	17.68546
20	343	367.8787	−24.87872

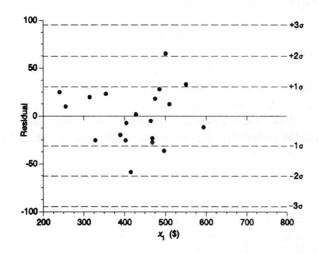

Multiple Regression

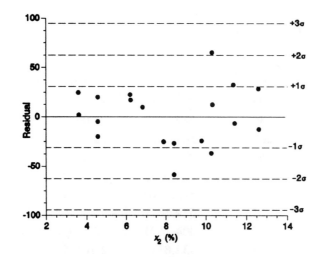

Solution a. The residual for the first y value is calculated as follows:

Residual = observed y value − estimated mean
= $y - \hat{y}$ = 405 − 382.03 = 22.97

For the second y value the residual is:

Residual = observed y value − estimated mean
= $y - \hat{y}$ = 285 − 259.93 = 25.07

These residuals agree (after rounding) with the first two entries in the third column labeled Residual in the accompanying table. Similar calculations produce the remaining residuals.

b. In both residual plots shown here, the residuals appear to be randomly distributed around the 0 line, as expected. If either of the plots had exhibited a mound shape or a bowl shape, this would have indicated that curvature should be added to the model.

EXAMPLE 14.11 The data for the electric bill prediction example used throughout this chapter are reproduced in the table with one important difference: the third value of the dependent variable, $607, has been changed to $407. The model

$$y = \beta_0 + \beta_1 x_1 + \beta_2 x_2 + \epsilon$$

is then fit to these (modified) data, with the printout as shown below. The residuals are also shown and plotted against the independent variable x_1, the total amount billed in the current year. Analyze the residual plot.

Name:	y($)	X1($)	X2(%)
1	405	355	6.2
2	285	240	3.6
3	407	500	10.3
4	468	465	4.6
5	627	550	11.4
6	572	487	12.6
7	503	498	10.3
8	385	390	4.6
9	435	428	3.6
10	410	401	7.9
11	393	415	8.4
12	630	594	12.6
13	473	468	8.4
14	355	315	4.6
15	453	404	11.4
16	485	468	9.8
17	304	255	6.8
18	563	510	10.3
19	511	476	6.2
20	343	328	7.9

Variable	Coefficient	Standard Error	t-test	.0500 Signif	p-value
a	42.8021	42.33377	1.001	N	.326
X1($)	.8635764	.1280934	6.742	Y	.000
X2(%)	4.767619	4.048048	1.178	N	.255

The critical t-value for d.f.= 17 and alpha= .0500 is 2.1098

Std dev of reg= 39.34283   R= .925   R-Sqrd= .856   Adj R-Sqrd= .839
F-test= 50.351   p-value= .0000   F-value from table (alpha= .05)= 3.59
Observations= 20.   Degrees of freedom for numerator= 2, for denominator= 17

Obs.	Actual y ($)	Predicted y ($)	Residual
1	405	378.9310	26.06903
2	285	267.2239	17.77612
3	407	523.6968	−116.6968
4	468	466.2962	1.703796
5	627	572.1199	54.88007
6	572	523.4358	48.56421
7	503	521.9697	−18.96967
8	385	401.5280	−16.52795
9	435	429.5762	5.423767
10	410	426.7604	−16.76044
11	393	441.2343	−48.23431
12	630	615.8385	14.16150
13	473	487.0039	−14.00388
14	355	336.7597	18.24026
15	453	446.0378	6.962158
16	485	493.6785	−8.678528
17	304	295.4339	8.566101
18	563	532.3326	30.66742
19	511	483.4237	27.57629
20	343	363.7194	−20.71936

Multiple Regression

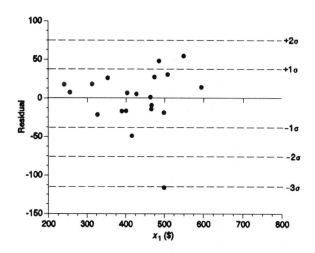

Solution The residual plot readily indicates the altered measurement at $x_1 = \$500$. The residual at this value of x_1 lies almost 3 standard deviations below 0, while no other residual lies more than about 1.5 standard deviations from 0. There are several possible reasons for the presence of outliers. Data may have been entered incorrectly, or the measurement may not have been taken under the conditions you are attempting to model. For example, the lower than expected value of the dependent variable may be due to a one-time discount because the homeowner made improvements to make the structure more energy-efficient. In many instances, there may be no obvious explanation for the presence of the outlier. You may want to rerun the regression analysis with the outlier omitted from the data to determine its effect on the analysis. Below is a new plot of the residuals (against x_1) when the regression is run without the outlier. Notice that only one residual is as much as 2 standard deviations away from 0 and most are within 1 standard deviation of 0. Note that the standard deviation has decreased from 39.3 with the outlier to 26.68 without the outlier, indicating a model that will provide more precise estimates and predictions.

Variable	Coefficient	Standard Error	t-test	.0500 Signif	p-value
a	26.04615	28.94674	.900	N	.382
X1($)	.8994517	.08723658	10.310	Y	.000
X2(%)	5.735685	2.753861	2.083	N	.054

The critical t-value for d.f.= 16 and alpha= .0500 is 2.1199

Std dev of reg= 26.68562   R= .968   R-Sqrd= .937   Adj R-Sqrd= .929
F-test= 118.53   p-value= .0000   F-value from table (alpha = .05)= 3.63
Observations= 19. Degrees of freedom for numerator= 2, for denominator= 16

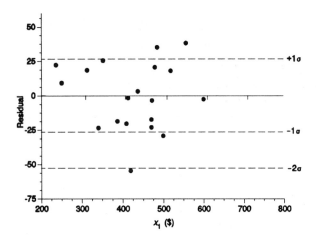

EXERCISES

14.8 Consider the executive salary and experience data in the table below. Log y is related to years of experience, x, using the second-order model

$$\log y = \beta_0 + \beta_1 x + \beta_2 x^2 + \epsilon$$

The regression output and a plot of the residuals [in terms of numbers of standard deviations (positive or negative)] from the mean 0 are shown below. Does the assumption of constant variance appear reasonable?

Years of Experience x	Salary y	Years of Experience x	Salary y	Years of Experience x	Salary y
7	$ 52,150	21	$ 87,256	28	$198,278
28	158,740	4	32,210	23	105,248
23	131,452	24	131,288	17	101,188
18	83,966	20	126,044	25	106,544
19	124,618	20	95,560	26	130,686
15	82,308	15	77,706	19	92,432
24	107,220	25	133,074	16	108,576
13	67,394	25	134,894	3	41,688
2	44,888	28	129,570	12	65,172
8	65,124	26	123,162	23	142,470
20	86,152	27	141,356	20	73,060
21	112,000	20	102,602	19	105,490
18	117,334	18	78,692	27	134,564
7	44,420	1	49,666	25	161,862
2	41,042	26	131,858	12	64,606
18	99,454	20	83,442	11	76,742
11	66,466	26	165,282		

The regression equation is LOGY = 10.54 + 0.0497 X + 0.000009 XX

Predictor	Coef	Stdev	t-ratio
Constant	10.53604	0.08479	124.26
X	0.04969	0.01182	4.20
XX	0.0000093	0.0003753	0.02

s = 0.1557   R-sq = 86.4%   R-sq(adj) = 85.8%

Analysis of Variance

SOURCE	DF	SS	MS
Regression	2	7.2119	3.6059
Error	47	1.1400	0.0243
Total	49	8.3519	

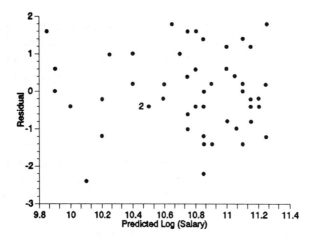

14.9 Refer to Exercise 14.8. A stem-and-leaf display of the residuals is given here. Analyze the display, especially with regard to the normality assumption.

Stem	Leaf
−.3	3, 5
−.2	0, 1, 1
−.1	0, 1, 2, 5, 6, 6, 7, 8
−.0	0, 1, 2, 2, 3, 5, 5, 5, 5, 5, 6, 7, 7, 8
.0	1, 2, 3, 3, 4, 4, 5, 7, 8, 8
.1	0, 4, 5, 6, 8, 8
.2	1, 1, 3, 4, 5, 6, 6

CHAPTER FIFTEEN

Model Building

Summary

This chapter discussed the selection of an appropriate model for a given set of data. In a **model building** effort, the researcher will utilize knowledge of the process being modeled, in addition to formal statistical procedures.

The first step in the construction of a model for a response variable y is the identification of a set of independent variables, each of which is classified as either **qualitative** or **quantitative**. If the number of potentially important independent variables is very large, a **stepwise regression** procedure may be employed to reduce the set by screening out those variables which seem unimportant for the prediction of y.

The researcher is advised to consider at least **second-order models**, those which contain **quadratic terms** and **two-way interactions** among the quantitative variables. The model may be modified and improved by testing sets of β parameters for significance.

The chapter concluded with a discussion of problems which may arise in constructing a prediction model.

15.1 The Two Types of Independent Variables: Quantitative and Qualitative

EXAMPLE 15.1 Most automobile loan applications request biographical and demographic information on such variables as:

a. Gender   b. Marital status

c. Age   d. Monthly income

e. Amount of loan desired   f. Credit references

For each of these variables, specify its type (qualitative or quantitative) and describe the nature of the levels that you might observe.

Solution

a. The variable for gender is qualitative, since its levels, "male" and "female," are not numerical, but instead are descriptive labels.

b. The levels of marital status (married, single, divorced, separated, widowed) are nonnumerical; hence, this is a qualitative variable.

c. Age is a quantitative variable with levels ranging from approximately 16 years to 100 years. (We assume that only those individuals of driving age would be applying for an automobile loan.)

Model Building

d. This variable assumes numerical values within a very large range representing all possible values of an individual's monthly income, and thus is quantitative.

e. The amount of the loan requested is a quantitative variable which will assume a numerical value between $0 and $10,000, say.

f. The levels will be nonnumerical labels representing, for example, names of banks or major credit cards; thus, credit reference is a qualitative variable.

EXERCISE

15.1 For each of the following variables, state its type and describe the nature of the levels which may be observed.

a. Social status

b. Percentage of unemployed persons in Montana on a specified date

c. Proportion of grand jury members who have attained a college degree

d. Occupation

e. Distance from residence to place of employment

15.2 Models with a Single Quantitative Independent Variable

EXAMPLE 15.2 Records for nine individuals who have purchased term life insurance policies with a particular company yielded the following information on age (in years) and amount of monthly premium (in dollars):

Age x	Monthly Premium y
20	3.86
25	4.03
30	4.22
35	4.50
40	4.94
45	5.50
50	6.26
55	7.44
60	9.03

a. Plot the data on a scattergram.

b. Suggest a model to relate $E(y)$ and x.

Solution a.

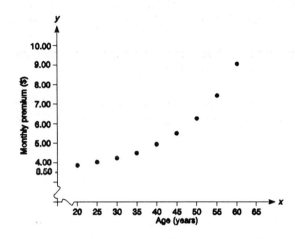

b. It is clear from the scattergram that the straight-line model

$$E(y) = \beta_0 + \beta_1 x$$

is inappropriate, as the relationship between y and x appears to be linear only over the restricted age range from 20–40 years. In the upper age range, the amount of increase in premium for a unit (one-year) increase in age increases as age increases. That is, the **rate** of increase in $E(y)$ is believed to increase as x increases. Thus, we may wish to fit a quadratic (second-order polynomial) model to the data:

$$E(y) = \beta_0 + \beta_1 x + \beta_2 x^2$$

where β_2 is expected to be positive to reflect the upward curvature of the response curve.

EXAMPLE 15.3 Refer to Example 15.2. The least squares quadratic prediction equation was determined to be

$$\hat{y} = 6.49 - .2x + .004x^2$$

with $s_{\hat{\beta}_2} = .0006$.

Test the hypothesis that the monthly premium increases at an increasing rate with age. Use a significance level of $\alpha = .05$.

Solution The parameter of interest is β_2, and the test has the following elements:

H_0: $\beta_2 = 0$
H_a: $\beta_2 > 0$

At $\alpha = .05$, the null hypothesis will be rejected if

$$t > t_{.05} = 1.943$$

where t is based on $n - (k + 1) = 9 - (2 + 1) = 6$ degrees of freedom.

Model Building

The test statistic is

$$t = \frac{\hat{\beta}_2}{s_{\hat{\beta}_2}} = \frac{.004}{.0006} = 6.67$$

This computed value lies within the rejection region. We thus conclude that β_2 is significantly larger than zero, i.e., that the amount of monthly premium increases at an increasing rate with age.

EXERCISE

15.2 A company which sells replacement computer parts advertises that it ships orders immediately upon receipt. The management is interested in modeling the relationship between time required for delivery of a customer's order, y, and the distance between shipping point and destination, x. It is believed that the model should allow for the average delivery time to increase as the distance increases to a certain value (say, 1000 miles), after which the amount of increase in delivery time for a unit increase in distance will decrease.

   a. Write a model to relate $E(y)$ and x.

   b. Sketch the shape of the response curve for the hypothesized model.

15.3 Models with Two or More Quantitative Independent Variables

EXAMPLE 15.4 Researchers for a consumer agency wish to investigate the relationship between mileage per gallon ratings (y), and the independent variables engine size (x_1, in cubic inches) and speed of automobile (x_2, in miles per hour).

   a. Write the first-order model for $E(y)$. What are the assumptions implied by the model?

   b. Write the first-order model plus interaction for $E(y)$. Graph $E(y)$ versus x_2 as you would expect it to appear for $x_1 = 250$ and $x_1 = 400$, where x_2 ranges from 25 miles per hour to 60 miles per hour.

   c. Write the complete second-order model for $E(y)$ as a function of x_1 and x_2.

Solution

   a. The general first-order model for $k = 2$ quantitative independent variables is:

$$E(y) = \beta_0 + \beta_1 x_1 + \beta_2 x_2$$

This model assumes that there is no curvature in the response curve and that the variables engine size (x_1) and speed of automobile (x_2) affect mileage per gallon ratings (y) independently of each other. All contour lines will be parallel.

b. The first-order model with interaction is given by

$$E(y) = \beta_0 + \beta_1 x_1 + \beta_2 x_2 + \beta_3 x_1 x_2$$

The inclusion of the interaction term $\beta_3 x_1 x_2$ allows contour lines to be nonparallel.

For $x_1 = 250$,

$$E(y) = \beta_0 + \beta_1(250) + \beta_2 x_2 + \beta_3(250)x_2$$
$$= (\beta_0 + 250\beta_1) + (\beta_2 + 250\beta_3)x_2$$

and for $x_1 = 400$,

$$E(y) = (\beta_0 + 400\beta_1) + (\beta_2 + 400\beta_3)x_2$$

Note the differences in the slopes and y-intercepts of the contour lines as the size of the automobile engine changes.

Contour lines for $x_1 = 250$ and $x_1 = 400$ may graph as follows:

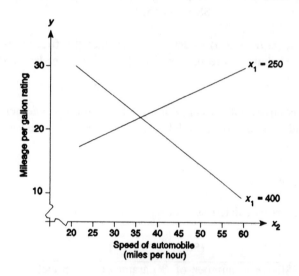

c. To allow for curvature in the contour lines, we may add quadratic terms to the model of part b to obtain the complete second-order model:

$$E(y) = \beta_0 + \beta_1 x_1 + \beta_2 x_2 + \beta_3 x_1 x_2 + \beta_4 x_1^2 + \beta_5 x_2^2$$

EXERCISE

15.3 A large grocery store chain is interested in predicting a family's weekly food expenditure (y) based on weekly income (x_1) and the number of people in the family (x_2).

a. Write the first-order model for $E(y)$. Interpret the model in terms of the problem.

Model Building

b. Do you think an interaction term would be appropriate for the model? Why or why not? Add an interaction term to the model of part a.

c. Write the complete second-order model for $E(y)$.

15.4 Model Building: Testing Portions of a Model

EXAMPLE 15.5 Refer to Example 15.4. The three models formulated to describe the relationship between mileage per gallon ratings (y), and the independent variables engine size (x_1) and speed of automobile (x_2), were fit to sample data for $n = 50$ automobiles with the following results:

(1) First-order model: $E(y) = \beta_0 + \beta_1 x_1 + \beta_2 x_2$
$SSE_1 = 80.4$

(2) First-order model with interaction: $E(y) = \beta_0 + \beta_1 x_1 + \beta_2 x_2 + \beta_3 x_1 x_2$
$SSE_2 = 62.9$

(3) Second-order model: $E(y) = \beta_0 + \beta_1 x_1 + \beta_2 x_2 + \beta_3 x_1 x_2 + \beta_4 x_1^2 + \beta_5 x_2^2$
$SSE_3 = 55.8$

Perform a test (at $\alpha = .05$) to determine whether the first-order model with interaction contributes more information than the first-order model for the prediction of y.

Solution We wish to compare model (1) with model (2) and test whether the interaction term $\beta_3 x_1 x_2$ should be retained in model (2). The appropriate test is then

H_0: $\beta_3 = 0$
H_a: $\beta_3 \neq 0$

The F statistic is calculated as follows:

$$F = \frac{(SSE_1 - SSE_2)/1}{SSE_2/[n - \text{number of } \beta \text{ parameters in model (2)}]}$$

and is based on $\nu_1 = 1$ numerator and $\nu_2 = 50 - 4 = 46$ denominator degrees of freedom. At significance level $\alpha = .05$, we will reject H_0 if

$F > F_{.05} \approx 4.06$

Substitution of the given values of SSE for the two models yields:

$$F = \frac{(80.4 - 62.9)/1}{62.9/(50 - 4)} = 12.80$$

This value of the test statistic lies within the rejection region. We thus conclude that the first-order model with interaction provides more information than the first-order model for predicting y; the interaction term $\beta_3 x_1 x_2$ makes an important contribution to the model.

EXAMPLE 15.6 Refer to Examples 15.4 and 15.5. Perform a test to determine whether the second-order model contributes more information than the first-order model with interaction for the prediction of y.

Solution To determine whether the second-order terms are useful in the model, we will compare model (2) with model (3) and perform a test of

$$H_0: \beta_4 = \beta_5 = 0$$
$$H_a: \text{At least one of the coefficients, } \beta_4 \text{ or } \beta_5, \text{ is nonzero.}$$

The test statistic is of the form

$$\frac{(SSE_2 - SSE_3)/2}{SSE_3/[n - \text{number of } \beta \text{ parameters in model (3)}]}$$

based on $\nu_1 = 2$ numerator degrees of freedom (because there are 2 parameters specified in H_0) and $\nu_2 = 50 - 6 = 44$ denominator degrees of freedom. Thus, we will reject H_0 (at $\alpha = .05$) if

$$F > F_{.05} \approx 3.21$$

The test statistic is computed as follows:

$$F = \frac{(62.9 - 55.8)/2}{55.8/(50 - 6)} = 2.80$$

Since this value of the test statistic does not fall within the rejection region, we cannot reject at H_0 at $\alpha = .05$. There is insufficient evidence to conclude that the second-order model is more useful than the first-order model with interaction for predicting mileage per gallon ratings.

EXERCISES

15.4 Refer to Exercise 15.3. The three models relating weekly food expenditure (y) to weekly income (x_1) and family size (x_2) were fit to $n = 80$ observations with the following results:

First-order model: SSE = 576.34
First-order model with interaction: SSE = 382.19
Second-order model: SSE = 104.65

Perform a test (at $\alpha = .05$) to determine whether the interaction term is a useful addition to the first-order model.

Model Building

15.5 Refer to Exercises 15.3 and 15.4. Perform a test to determine whether the second-order model contributes more information than the first-order model with interaction for predicting weekly food expenditure. Use $\alpha = .05$.

15.5 Models with One Qualitative Independent Variable

EXAMPLE 15.7 A management consulting firm is interested in constructing a model for the annual salary of bank managers, and believes that the manager's gender is an important independent variable to consider.

  a. Write a model which will provide a single prediction equation for the mean salary of male and female managers, if gender is the only independent variable of interest.

  b. Interpret the parameters of the model in part a.

Solution a. Let μ_M be the mean salary of a male bank manager and μ_F the mean salary of a female manager. We can model $E(y)$, the mean annual salary, as follows:

$$E(y) = \beta_0 + \beta_1 x_1$$

where $x_1 = \begin{cases} 1 & \text{if the manager is male} \\ 0 & \text{if the manager is female} \end{cases}$

  b. To determine the mean annual salary for a male bank manager, we let the dummy variable x_1 assume the value 1. Then

$$\mu_M = E(y) = \beta_0 + \beta_1(1) = \beta_0 + \beta_1$$

Similarly, for a female bank manager,

$$\mu_F = E(y) = \beta_0 + \beta_1(0) = \beta_0$$

Thus, β_0 is the mean salary of a female bank manager and $\beta_1 = \mu_M - \mu_F$ is the difference in the mean salaries for male and female bank managers.

EXAMPLE 15.8 The owner of a fast-food chain with five locations (L_1, L_2, L_3, L_4, and L_5) wishes to model his daily sales as a function of location.

  a. Write a model for mean daily sales, $E(y)$, as a function of location.

  b. Interpret the parameters of the model in part a.

  c. What is the difference (in terms of the model parameters) between the mean daily sales for locations L_3 and L_1?

  d. What is the difference (in terms of the model parameters) between the mean daily sales for locations L_2 and L_5?

Solution

a. The model relating $E(y)$ to the qualitative variable location is

$$E(y) = \beta_0 + \beta_1 x_1 + \beta_2 x_2 + \beta_3 x_3 + \beta_4 x_4$$

where $x_1 = \begin{cases} 1, & \text{if Location } L_2 \\ 0, & \text{otherwise} \end{cases}$ $\quad x_2 = \begin{cases} 1, & \text{if Location } L_3 \\ 0, & \text{otherwise} \end{cases}$

$x_3 = \begin{cases} 1, & \text{if Location } L_4 \\ 0, & \text{otherwise} \end{cases}$ $\quad x_4 = \begin{cases} 1, & \text{if Location } L_5 \\ 0, & \text{otherwise} \end{cases}$

Note that four dummy variables are required to describe the five levels of the qualitative independent variable.

b. The parameter β_0 represents the mean daily sales at location L_1, the base level of the qualitative independent variable. The remaining parameters represent the differences between mean daily sales for the particular location and location L_1. Thus, for example, β_3 is the difference between the mean daily sales at location L_4 and location L_1. This can also be seen by substituting the appropriate values of the dummy variables into the model for $E(y)$:

For location L_4 (set $x_1 = 0, x_2 = 0, x_3 = 1, x_4 = 0$):

$$\mu_4 = E(y) = \beta_0 + \beta_1(0) + \beta_2(0) + \beta_3(1) + \beta_4(0)$$
$$= \beta_0 + \beta_3$$

For location L_1 (set $x_1 = x_2 = x_3 = x_4 = 0$):

$$\mu_1 = E(y) = \beta_0 + \beta_1(0) + \beta_2(0) + \beta_3(0) + \beta_4(0)$$
$$= \beta_0$$

Thus, $\mu_4 - \mu_1 = \beta_3$, as obtained above. The parameters β_1, β_2, and β_4 have analogous interpretations.

c. Location L_1 is the base level of the qualitative independent variable; thus, the difference between mean daily sales for locations L_3 and L_1 is the parameter representing level L_3, namely, β_2:

$$\beta_2 = \mu_3 - \mu_1$$

where μ_3 and μ_1 are the mean daily sales at locations L_3 and L_1, respectively.

d. To model the mean daily sales for location L_2, set $x_1 = 1, x_2 = x_3 = x_4 = 0$:

$$\mu_2 = \beta_0 + \beta_1(1) + \beta_2(0) + \beta_3(0) + \beta_4(0)$$
$$= \beta_0 + \beta_1$$

To model the mean daily sales for location L_5, set $x_1 = x_2 = x_3 = 0, x_4 = 1$:

$$\mu_5 = \beta_0 + \beta_1(0) + \beta_2(0) + \beta_3(0) + \beta_4(1)$$
$$= \beta_0 + \beta_4$$

Model Building

The difference between the mean daily sales for locations L_2 and L_5 is

$$\mu_2 - \mu_5 = (\beta_0 + \beta_1) - (\beta_0 + \beta_4)$$
$$= \beta_1 - \beta_4$$

EXERCISE

15.6 A company wished to determine how well its management trainee program predicted an individual's success five years after completion of the program. For graduates of the training program, they chose to model the mean salary, $E(y)$, as a function of the qualitative independent variable performance in the program (excellent, fair, or poor).

a. Define dummy variables to describe the levels of the qualitative independent variable.

b. Write the model relating $E(y)$ to performance in the training program.

c. Based on the model in part b, what is the mean salary for employees who were rated excellent?

d. In terms of the model parameters, what is the difference between the mean salaries for employees who were rated poor and those who were rated excellent?

15.6 Comparing the Slopes of Two or More Lines
15.7 Comparing Two or More Response Curves

Note: In Examples 15.9–15.19, we will proceed to build a model in stages. Graphical interpretations will be provided at each stage.

EXAMPLE 15.9 Suppose we wish to relate $E(y)$, the mean salary of a bank manager, to years of experience for bank managers in four cities (Chicago, New York, Miami, and St. Louis). Write a model for $E(y)$ which allows for a single straight-line relationship between mean salary and years of experience for all four cities. Graph a typical response curve.

Solution The straight-line model

$$E(y) = \beta_0 + \beta_1 x_1$$

where x_1 = years of experience, is unable to detect a difference in mean salaries of bank managers among the four cities, if such differences exist. According to this model, a single straight line characterizes the relationship between mean salary and years of experience, as shown in the figure.

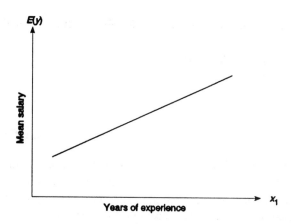

EXAMPLE 15.10 Refer to Example 15.9. Write a model which will allow the straight lines relating mean salary $E(y)$ to years of experience x_1 to differ from one city to another, but in such a manner that the increase in mean salary per year increase in experience is the same for the four cities.

Solution A model which allows the response lines to be parallel, but with different y-intercepts, is

$$E(y) = \beta_0 + \beta_1 x_1 + \beta_2 x_2 + \beta_3 x_3 + \beta_4 x_4$$

where x_1 = years of experience

$x_2 = \begin{cases} 1, \text{ if New York} \\ 0, \text{ otherwise} \end{cases}$

$x_3 = \begin{cases} 1, \text{ if Miami} \\ 0, \text{ otherwise} \end{cases}$

$x_4 = \begin{cases} 1, \text{ if St. Louis} \\ 0, \text{ otherwise} \end{cases}$

Typical response curves are shown in the figure:

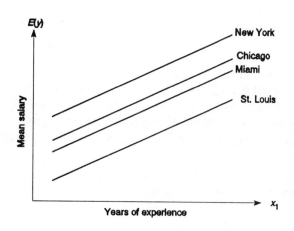

EXAMPLE 15.11 Refer to Examples 15.9 and 15.10. Write a model which allows the straight lines relating mean salary $E(y)$ to years of experience x_1 to differ for the four cities.

Model Building

Solution The model which allows the four straight lines to have different y-intercepts and slopes is obtained by adding interaction terms to the model developed in Example 15.10:

$$E(y) = \beta_0 + \beta_1 x_1 + \beta_2 x_2 + \beta_3 x_3 + \beta_4 x_4 + \beta_5 x_1 x_2 + \beta_6 x_1 x_3 + \beta_7 x_1 x_4$$

This model permits nonparallel response lines, as shown in the figure.

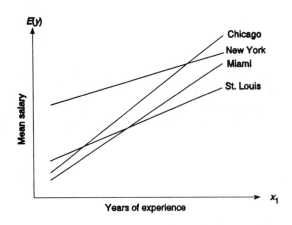

EXAMPLE 15.12 Refer to Example 15.11. Explain how you would perform a test to determine whether the four response lines differ (as in Example 15.11 figure) or whether a single line characterizes mean salary for all four cities (as in Example 15.9 figure).

Solution We wish to compare the following models:

Model (1): $E(y) = \beta_0 + \beta_1 x_1$

Model (3): $E(y) = \beta_0 + \beta_1 x_1 + \beta_2 x_2 + \beta_3 x_3 + \beta_4 x_4 + \beta_5 x_1 x_2 + \beta_6 x_1 x_3 + \beta_7 x_1 x_4$

To determine whether model (3) provides more information than model (1) for predicting y, we perform a test of

H_0: $\beta_2 = \beta_3 = \beta_4 = \beta_5 = \beta_6 = \beta_7 = 0$
H_a: At least one of the coefficients, $\beta_2, \beta_3, \ldots, \beta_7$, is nonzero

If we let SSE_1 and SSE_3 denote the sum of squared errors for models (1) and (3), respectively, then we write the test statistic as

$$F = \frac{(SSE_1 - SSE_3)/6}{SSE_3/(n-8)}$$

The test statistic is based on an F distribution with $\nu_1 = 6$ numerator degrees of freedom (because there are 6 parameters specified in H_0) and $\nu_2 = (n-8)$ denominator degrees of freedom (because model (3) contains 8 β parameters).

Rejection of the null hypothesis implies that model (3) contributes more information than does model (1) for the prediction of mean salary.

EXAMPLE 15.13 Refer to Example 15.11. Explain how you would perform a test to determine whether there is evidence that years of experience and city interact to affect mean salary.

Solution It is required to fit the following two models, calculate the drop in the sum of squares for error, and conduct an F test:

Model (2): $\quad E(y) = \beta_0 + \beta_1 x_1 + \beta_2 x_2 + \beta_3 x_3 + \beta_4 x_4$

Model (3): $\quad E(y) = \beta_0 + \beta_1 x_1 + \beta_2 x_2 + \beta_3 x_3 + \beta_4 x_4 + \beta_5 x_1 x_2 + \beta_6 x_1 x_3 + \beta_7 x_1 x_4$

The relevant test is composed of the following elements:

H_0: $\beta_5 = \beta_6 = \beta_7 = 0$
H_a: At least one of the coefficients, β_5, β_6, or β_7, is nonzero

Test statistic:

$$F = \frac{(SSE_2 - SSE_3)/3}{SSE_3/(n - 8)}$$

where the distribution of F is based on $\nu_1 = 3$ numerator degrees of freedom and $\nu_2 = (n - 8)$ denominator degrees of freedom.

Rejection of the null hypothesis allows us to conclude that the independent variables years of experience and city interact to affect salary.

EXAMPLE 15.14 Suppose we believe that the relationship between mean salary $E(y)$ and years of experience x_1 is second-order. Write a model which allows for identical mean salary curves for all four cities.

Solution The following model yields a single second-order curve to describe the relationship between $E(y)$ and x_1 for the four cities.

$$E(y) = \beta_0 + \beta_1 x_1 + \beta_2 x_1^2$$

We would expect $E(y)$ to increase as the number of years of experience increases. However, the **rate** of increase would probably decrease as experience increases. A typical response curve might appear as follows:

Model Building

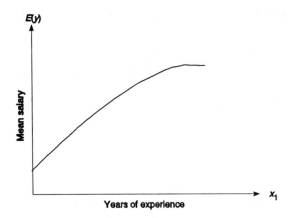

EXAMPLE 15.15 Add appropriate terms to the model of Example 15.14 to permit the response curves to have the same shapes, but different y-intercepts.

Solution We now add three dummy variables to represent the four levels of the qualitative independent variable city:

$$E(y) = \beta_0 + \beta_1 x_1 + \beta_2 x_1^2 + \beta_3 x_2 + \beta_4 x_3 + \beta_5 x_4$$

where x_2, x_3, and x_4 are as defined in Example 15.10. According to this model, the second-order response curves for the four cities have the same shape, but different y-intercepts, as shown in the figure:

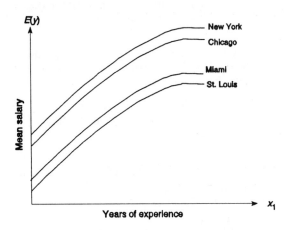

EXAMPLE 15.16 Add appropriate terms to the model of Example 15.15 to allow different response curves for the four cities.

Solution We now add terms representing interaction between years of experience and city:

$$E(y) = \beta_0 + \beta_1 x_1 + \beta_2 x_1^2 + \beta_3 x_2 + \beta_4 x_3 + \beta_5 x_4 + \beta_6 x_1 x_2 + \beta_7 x_1 x_3 \\ + \beta_8 x_1 x_4 + \beta_9 x_1^2 x_2 + \beta_{10} x_1^2 x_3 + \beta_{11} x_1^2 x_4$$

This model yields typical response curves as shown in the figure:

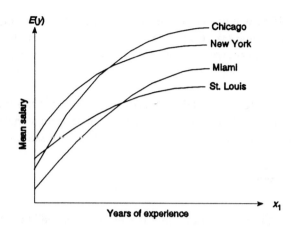

EXAMPLE 15.17 Refer to Example 15.16. Give the equation of the second-order model for the mean salary of bank managers in New York.

Solution We substitute $x_2 = 1$, $x_3 = 0$, and $x_4 = 0$ into the second-order model:

$$E(y) = \beta_0 + \beta_1 x_1 + \beta_2 x_1^2 + \beta_3(1) + \beta_4(0) + \beta_5(0) + \beta_6 x_1(1) + \beta_7 x_1(0)$$
$$+ \beta_8 x_1(0) + \beta_9 x_1^2(1) + \beta_{10} x_1^2(0) + \beta_{11} x_1^2(0)$$
$$= (\beta_0 + \beta_3) + (\beta_1 + \beta_6)x_1 + (\beta_2 + \beta_9)x_1^2$$

EXAMPLE 15.18 Specify the null hypothesis which would be tested to determine whether the second-order curves are identical for the four cities.

Solution If the four response curves are in fact identical, it is not necessary to include the independent variable city in the model; thus, all terms involving x_2, x_3, or x_4 would be deleted, producing the model

$$E(y) = \beta_0 + \beta_1 x_1 + \beta_2 x_1^2$$

The appropriate null hypothesis would then be

$$H_0: \beta_3 = \beta_4 = \beta_5 = \cdots = \beta_{11} = 0$$

EXAMPLE 15.19 Suppose it is known that the response curves differ for the four cities, but we wish to determine whether the second-order terms contribute useful information for the prediction of y. Specify the appropriate null hypothesis which would be tested.

Solution If the second-order terms are not useful in the model, we would delete all terms involving x_1^2 and the null hypothesis would be:

$$H_0: \beta_2 = \beta_9 = \beta_{10} = \beta_{11} = 0$$

Model Building

EXERCISE

15.7 The Internal Revenue Service (IRS) would like to construct a model for $E(y)$, the mean computation error on submitted income tax returns. Write a straight-line model for the relationship between mean computation error and annual gross income. Graph a typical response curve.

15.8 Refer to Exercise 15.7. Add terms to the model for $E(y)$ which represent the levels for type of return submitted ("long" form or "short" form). Assume the independent variables, annual gross income and type of form submitted, do not interact to affect the mean computation error. Graph the response curves.

15.9 Refer to Exercise 15.8.

    a. Now add interaction terms to the model to allow the response curves to differ in slope and y-intercept. Graph typical response curves.

    b. Specify the hypothesis you would test to determine whether there is a difference in the mean computation error for long forms and short forms.

15.10 Refer to Exercises 15.7–15.9. Suppose we believe that the relationship between mean computation error $E(y)$ and annual gross income is second-order. Write a model which allows for identical response curves for the two types of forms. Graph a typical response curve.

15.11 Refer to Exercise 15.10. Add appropriate terms to the model to permit the two response curves to have the same shape, but different y-intercepts. Graph typical response curves.

15.12 Add appropriate terms to the model of Exercise 15.11 to allow for interaction between annual gross income and type of form submitted. Graph typical response curves.

15.13 Refer to Exercise 15.12. Give the equation of the second-order model for the mean computation error on long forms.

15.14 Refer to Exercise 15.12. Specify the null hypothesis which would be tested to determine whether the second-order curves are identical for the two types of forms.

15.15 Refer to Exercise 15.12. Suppose it is known that the response curves differ for the two types of forms, but we wish to determine whether the second-order terms make a useful contribution to the model. Specify the null hypothesis which would be tested.

15.8 Model Building: Stepwise Regression

EXAMPLE 15.20 The rate of unemployment in the United States is often used to describe the state of the economy. Following is a list of variables which are thought to influence $y =$ the monthly unemployment rate:

$x_1 =$ inflation rate
$x_2 =$ price per ounce of gold
$x_3 =$ Dow Jones Industrial closing value (monthly average)
$x_4 =$ prime interest rate
$x_5 = \begin{cases} 1 & \text{if month between October and March} \\ 0 & \text{if month between April and September} \end{cases}$

Data from two previous years are shown below. Use a stepwise regression procedure to select from the above list the variables that should be included in a model for unemployment rate.

Unemployment Rate, y	Inflation Rate, x_1	Price per Ounce of Gold, x_2	Dow Jones Average, x_3	Prime Interest Rate, x_4	Dummy Variable for Month, x_5
5.9	13.0	250	870	12.0	1
6.0	12.1	250	840	12.1	1
6.0	14.0	250	880	12.1	1
5.7	12.5	245	830	12.1	1
5.6	12.0	245	840	12.0	0
5.6	12.0	240	840	12.0	0
5.5	11.1	245	830	12.0	0
5.6	11.0	250	820	11.9	0
5.7	10.5	250	830	11.9	0
5.7	10.5	255	860	11.9	0
6.0	13.0	260	880	11.9	1
5.9	12.0	255	870	11.8	1
5.9	12.0	255	850	11.8	1
5.8	11.0	250	840	11.8	1
5.7	15.0	265	810	11.8	1
5.7	13.0	260	860	11.8	1
5.8	14.1	260	855	11.8	0
5.8	14.0	285	820	11.8	0
5.6	13.0	300	842	11.6	0
5.7	12.9	315	848	11.8	0
6.0	14.1	335	890	12.3	0
5.8	14.0	400	880	13.3	0
6.0	12.8	430	878	15.0	1
5.8	12.8	415	820	15.6	1

Model Building

Solution The SAS stepwise regression printout is presented below.

```
STEP 1        VARIABLE  X3  ENTERED           R-SQUARE = 0.406

              DF    SUM OF SQUARES   MEAN SQUARE    F     PROB > F
REGRESSION    1       0.21657          0.217      15.04   0.0008
ERROR        22       0.31676          0.0144
TOTAL        23       0.53333

                B VALUE    STD ERROR      F      PROB > F
INTERCEPT       2.185
X3              0.0042      0.00109     15.04     0.0008

STEP 2        VARIABLE  X5  ENTERED           R-SQUARE = 0.634

              DF    SUM OF SQUARES   MEAN SQUARE    F     PROB > F
REGRESSION    2       0.33791          0.169      18.16   0.0001
ERROR        21       0.19542          0.0093
TOTAL        23       0.53333

                B VALUE    STD ERROR      F      PROB > F
INTERCEPT       2.482
X3              0.0038      0.00089     18.40     0.0003
X5              0.144       0.03975     13.04     0.0017

STEP 3        VARIABLE  X1  ENTERED           R-SQUARE = 0.661

              DF    SUM OF SQUARES   MEAN SQUARE    F     PROB > F
REGRESSION    3       0.35268          0.118      13.02   0.0001
ERROR        20       0.18065          0.0090
TOTAL        23       0.53333

                B VALUE    STD ERROR      F      PROB > F
INTERCEPT       2.435
X1              0.0213      0.01669      1.64     .2155
X3              0.0035      0.00090     15.64     .0008
X5              0.1380      0.0394      12.27     .0022

STEP 4        VARIABLE  X1  REMOVED           R-SQUARE = 0.634

              DF    SUM OF SQUARES   MEAN SQUARE    F     PROB > F
REGRESSION    2       0.33791          0.169      18.16   0.0001
ERROR        21       0.19542          0.0093
TOTAL        23       0.53333

                B VALUE    STD ERROR      F      PROB > F
INTERCEPT       2.482
X3              0.0038      0.00089     18.40     0.0003
X5              0.144       0.03975     13.04     0.0017
```

The first variable included in the model is x_3, the Dow Jones average monthly closing. At the second step, the variable x_5, a dummy variable for month, enters the model. In the third step, x_1, the inflation rate, is selected for the model. However, the F statistic for x_1 ($F = 1.64$) is not statistically significant at the preassigned level of $\alpha = .10$; thus, x_1 is removed from the model in the next step. Since none of the other independent variables can meet the $\alpha = .10$ criterion for admission into the model, the stepwise procedure is terminated. We would then concentrate on the variables x_3 and x_5 in the effort to construct a model for $E(y)$. It remains to be decided how the variables should be entered into the prediction equation; the techniques described in Chapter 15 may be applied. (Note that the assumption of independent random errors is debatable for time series data such as these. One should be cautious in making inferences from the model.)

EXERCISE

15.16 Describe in detail a stepwise regression procedure; explain its importance as a method of choosing which of a large set of potential independent variables should be included in a model-building effort.

ANSWERS TO SELECTED EXERCISES

Chapter 1

1.1 the set of all 1,500 stocks listed on the NYSE (i.e., the population)

1.2 a. all 1993 college graduates with degrees in psychology
    b. a measure of the reliability of the inference

Chapter 2

2.1 a. nominal b. ratio c. nominal d. nominal
    e. ratio   f. ratio g. ratio

2.2 a. nominal b. nominal c. ratio d. ratio

2.8 $353.16

2.9 5.95 hours

2.10 8 days

2.11 3

2.14 $255

2.15 174.57, 13.21

2.16 370.11, 19.24

2.17 3,066.67, 55.38

2.18 42 seconds

2.19 a. It is possible that very few of the measurements will fall within (700, 1,700); at least 3/4 of the first-year sales will fall within the interval (200, 2,200).
    b. approximately 68%; approximately 95%
    c. approximately 2.5%; approximately 16%

2.20 $475; $91.67

2.21 a. $z = 2$ b. $z = -1$ c. 50th

Answers

2.22 a. $z = 2$ b. $z = -1$ c. $z = 0$
      d. $z = -.67$ e. $z = 4$

2.23 a. $z = -.5$ b. 17.36%

2.24 $z = 4$; very rare event based on past investment experience

2.25 $Q_L = 228$; median $= 266.5$; $Q_U = 291$

2.26 $Q_L = 171$; median $= 301.5$; $Q_U = 427$

2.27 a. 63

2.28 a. 256

Chapter 3

3.1   b. 3/10

3.2 $P(A) = .992$; $P(B) = .096$; $P(C) = .008$

3.3   a. .5   b. .5   c. 1   d. 0

3.4   a. $P(A) = .445$; $P(B) = .686$ b. $P(A \cup B) = .81$ c. $P(A \cap B) = .321$

3.5   a. A': {House has one bathroom}; $P(A') = .22$
      b. $A \cup B$: {House has two or three bedrooms *or* at least two bathrooms, or both};
          $P(A \cup B) = 1.0$
      c. .44   d. .34   e. .66

3.6   a. .83   b. .12   c. .42   d. .17   e. .88

3.7   a. .51   b. .75

3.8   a. .87   b. .90

3.9   a. 1/12; 2/3   b. no   c. no

3.10 a. yes   b. no   c. no

3.11 .24

3.12 b, e

3.13 a, c, d

3.15 a. 24   b. 210   c. 84   d. 84   e. 1

3.16 2184

3.17 495

3.18 1,680

3.19 112/120 ≈ .933

3.20 462/495 ≈ .933

Chapter 4

4.1 a. discrete; $x = 0, 1, 2, ...$ b. discrete; $x = 0, 1, 2, ...$
    c. continuous; $x > 0$ d. continuous; $x \geq 0$
    e. continuous; $x > 0$ f. discrete; $x = 0, 1, 2, ...$

4.3 a. .015 b. .985 c. .156

4.4 a.

x	0	1	2
$p(x)$	.7225	.2550	.0225

b. .2775

4.5 a. 2.04; 1.3584; 1.17 b. .98

4.7 a. no b. yes c. yes

4.8 a.

x	0	1	2	3
$p(x)$	.343	.441	.189	.027

b. .216

4.9 a. .787 b. .595 c. .783 d. 6; 3.6 e. .939

4.10 a. .351 b. .9933

4.11 a. .135 b. $(.5940)^2 = .353$

Chapter 5

5.1 a. 1/3 b. 2/9

5.2 a. .75 b. 50 minutes, 11.55 minutes

5.3 a. .0401 b. .0532 c. .0838 d. .7791

Answers

5.4 a. 2.58 b. −1.96 c. −.75

5.5 a. .4400 b. 1.22%

5.6 a. .7794 b. $2,983

5.7 .8365

5.8 .9599

5.9 a. .329680 b. .450852

5.10 .367879

Chapter 6

6.1 a.

m	2	3	4
p(m)	.3	.4	.3

b. $E(m) = 2(.3) + 3(.4) + 4(.3) = 3 = \mu$

6.2 a.

m	3	4	5
p(m)	.3	.4	.3

b. $E(m) = 4, \mu = 4.6$

6.3 $P(z \leq -2.13) = .0166$

6.4 a. approximately normal with mean $\mu_{\bar{x}} = \$41,000$ and standard deviation $\sigma_{\bar{x}} = \$3,000$
    b. 100   c. .8557, .9953

Chapter 7

7.1 a. 1.645 b. 2.33 c. 2.58

7.2 b. $429 ± $14.13

7.3 $19.80 ± $.67

7.4 67

7.5 a. −2.552 b. 1.895 c. 1.717

7.6 a. (499.37, 540.63)

7.7 a. 1,712 ± 110.9

7.8 $.64 \pm .124$

7.9   Since .60 is in the interval, do not cease home deliveries.

7.10   4157

Chapter 8

8.1 $z = -3$; yes

8.2 $z = 1.44$; no

8.3 p-value $= .0013$

8.4 p-value $= 14.98$

8.5 $t = 2.86$; yes

8.6   a. $t = -1.65$; no

8.7 $z = .83$; do not cease home deliveries

8.8 $z = 5.96$; yes

8.9 $\beta = .1492$ for $\alpha = .05$; $\beta = .3372$ for $\alpha = .01$

8.10   a.   29.1413   b.   9.39046

8.11   a.   989265   b.   5.00784

8.12 $.072 < \sigma^2 < .189$

8.13 $\chi^2 = 29.04$; no

Chapter 9

9.1 $z = 2.96$; yes

9.2 $\$45.00 \pm \$29.78$

9.3   a. $t = -.49$, do not reject H_0

9.4 -8.2 ± 37.30

9.5   a. $t = -1.21$; no

9.6   a. $\$1.18 \pm \$.55$

Answers

9.8 $z = 1.58$; no

9.9 $.06 \pm .062$

9.10 314, 314

9.11 77, 77

9.12 a. 4.94   b. 1.65   c. 2.90

9.13 $F = 1.59$; no

9.14 (.057, 13.434)

Chapter 10

10.1 a. independent sampling (completely randomized)   b. $F = 10.0$; yes
      c. 18.375 ± 6.404 d. -15.33 ± 12.68

10.2 a. $t = -1.05$; do not reject H_0 b. $F = 1.10$; do not reject H_0
      c. $t^2 = 1.10 = F$; $t_{.025}$ (8 df) = 2.306, $F_{.05}$ (1 df, 8 df) = 5.32 = $(2.306)^2$

10.3 μ_1 and μ_2 are not significantly different; μ_1 and μ_3 are not significantly different; μ_2 and μ_3 appear to be significantly different.

10.4 a. randomized block

      b.
Source	df	SS	MS	F
Restaurant (Treatments)	2	32.452	16.226	1.57
Day (Blocks)	4	54.307	13.577	1.31
Error	8	82.741	10.343	
Totals	14	169.500		

      c. $F = 1.57$; no   d. $F = 1.31$; no   e. 3.46 ± 3.78

10.5 a.
Source	df	SS	MS	F
Brand (Treatments)	2	38	19	9.5
Type (Blocks)	2	1,022	511	255.5
Error	4	8	2	
Totals	8	1,068		

      b. $F = 9.5$; yes   c. 3.00 ± 3.21

10.6 a.

Source	df	SS	MS	F
Union	1	7,168.444	7,168.444	26.34
Plan	2	5,425.167	2,712.584	9.97
Union-Plan Interaction	2	159.389	79.695	.29
Error	30	8,165.000	272.167	
Totals	35	20,918.000		

b. $F = .29$; no

10.7 a. (343.59, 365.75)   (Note: $t_{.025}$ with 30 degrees of freedom is approximately 1.645.)
b. (−40.67, −9.33)

Chapter 11

11.1 $S = 14$; reject H_0 ($p = .0577$)

11.2 b. $T_B = 47.5$; do not reject H_0

11.3 $T_- = 1.5$; reject H_0

11.4 b. $H = 11.65$; reject H_0

11.5 b. $F_r = 11.05$; reject H_0

11.6 a. .90   b. .90

11.7 reject H_0

Chapter 12

12.1 $X^2 = 43.83$; yes

12.2 $X^2 = 7.94$; no

12.3 $X^2 = 108.35$; reject H_0

12.4 $X^2 = 11.31$; no

Chapter 13

13.1 b. −2, 1.5

13.2 b. $\hat{y} = -15.20 + 191.34x$ c. 16,465.4   d. 616.22

Answers

13.3   2,058.175

13.4   16,465.4

13.5 $t = 5.86$; reject H_0

13.6   191.34 ± 75.324

13.7 $t = 1.50$; no

13.8   a. .90   b. .81

13.9 $t = 5.86$; yes

13.10 .219

13.11 673.62 ± 35.69

13.12 673.62 ± 91.62

13.14 a. $y = \beta_0 + \beta_1 x + \epsilon$ b. $y = 10.88 + .236x$ c. $t = 5.33$; reject H_0
       d. .936; .876   e. 15.596 ± 1.835

Chapter 14

14.1   a. $\hat{y} = 1.2093 + .3710x_1 + .0021x_2$ b. 3.91

14.2   a. .126   b. .021

14.3   a. $t = 2.82$; no

14.4   a. $y = \beta_0 + \beta_1 x_1 + \beta_2 x_2 + \epsilon$ b. $t = 2.67$; reject $H_0: \beta_2 = 0$

14.5   .7194

14.6 $F = 7.68$; reject $H_0: \beta_1 = \beta_2 = 0$

Chapter 15

15.1   a. qualitative (upper class, middle class, lower class)
       b. quantitative   c. quantitative
       d. qualitative   e. quantitative

15.2   a. $E(y) = \beta_0 + \beta_1 x_1 + \beta_2 x^2$

15.3 a. $E(y) = \beta_0 + \beta_1 x_1 + \beta_2 x_2$
 b. yes; $E(y) = \beta_0 + \beta_1 x_1 + \beta_2 x_2 + \beta_3 x_1 x_2$
 c. $E(y) = \beta_0 + \beta_1 x_1 + \beta_2 x_2 + \beta_3 x_1 x_2 + \beta_4 x_1^2 + \beta_5 x_2^2$

15.4 $F = 38.61$; yes

15.5 $F = 98.13$; yes

15.6 a. $x_1 = \begin{cases} 1, \text{ if fair performance} \\ 0, \text{ otherwise} \end{cases}$ $x_2 = \begin{cases} 1, \text{ if poor performance} \\ 0, \text{ otherwise} \end{cases}$
 b. $E(y) = \beta_0 + \beta_1 x_1 + \beta_2 x_2$ c. β_0 d. β_2

15.7 $E(y) = \beta_0 + \beta_1 x_1$, where $x_1 = $ annual gross income

15.8 $E(y) = \beta_0 + \beta_1 x_1 + \beta_2 x_2$, where $x_2 = \begin{cases} 1, \text{ if short form} \\ 0, \text{ otherwise} \end{cases}$

15.9 a. $E(y) = \beta_0 + \beta_1 x_1 + \beta_2 x_2 + \beta_3 x_1 x_2$ b. $H_0: \beta_2 = 0$

15.10 $E(y) = \beta_0 + \beta_1 x_1 + \beta_2 x_1^2$

15.11 $E(y) = \beta_0 + \beta_1 x_1 + \beta_2 x_1^2 + \beta_3 x_2$, where $x_2 = \begin{cases} 1, \text{ if short form} \\ 0, \text{ otherwise} \end{cases}$

15.12 $E(y) = \beta_0 + \beta_1 x_1 + \beta_2 x_1^2 + \beta_3 x_2 + \beta_4 x_1 x_2 + \beta_5 x_1^2 x_2$

15.13 $E(y) = \beta_0 + \beta_1 x_1 + \beta_2 x_1^2$

15.14 $H_0: \beta_3 = \beta_4 = \beta_5 = 0$

15.15 $H_0: \beta_2 = \beta_5 = 0$